CLIMATE CHANGE AND POPULATION HEALTH

A PRIMER

MONA SARFATY, MD, MPH, FAAFP

Executive Director
Medical Society Consortium on Climate and Health
Center for Climate Change Communication
George Mason University

JONES & BARTLETT
LEARNING

World Headquarters
Jones & Bartlett Learning
5 Wall Street
Burlington, MA 01803
978-443-5000
info@jblearning.com
www.jblearning.com

Jones & Bartlett Learning books and products are available through most bookstores and online booksellers. To contact Jones & Bartlett Learning directly, call 800-832-0034, fax 978-443-8000, or visit our website, www.jblearning.com.

Substantial discounts on bulk quantities of Jones & Bartlett Learning publications are available to corporations, professional associations, and other qualified organizations. For details and specific discount information, contact the special sales department at Jones & Bartlett Learning via the above contact information or send an email to specialsales@jblearning.com.

22360-6

Production Credits

VP, Product Management: Amanda Martin
Director of Product Management: Laura Pagluica
Product Manager: Sophie Fleck Teague
Content Strategist: Sara Bempkins
Project Manager: Kristen Rogers
Project Specialist: Kelly Sylvester
Senior Digital Project Specialist: Angela Dooley
Senior Marketing Manager: Susanne Walker
VP, Manufacturing and Inventory Control:
    Therese Connell

Composition: Exela Technologies
Project Management: Exela Technologies
Cover Design: Kristin E. Parker
Media Development Editor: Faith Brosnan
Rights and Permissions Manager: John Rusk
Rights Specialist: Liz Kincaid
Cover Image (Title Page): © BenGoode/
    Getty Images
Printing and Binding: McNaughton & Gunn

Library of Congress Cataloging-in-Publication Data

Names: Sarfaty, Mona, author.
Title: Climate change and population health : a primer / Mona Sarfaty, MD, MPH, FAAFP,
    Executive Director, Medical Society Consortium on Climate and Health,
    Center for Climate Change Communication, George Mason University.
Description: First edition. | Burlington : Jones & Bartlett Learning,
    [2022] | Includes bibliographical references and index.
Identifiers: LCCN 2020025786 | ISBN 9781284170207 (paperback)
Subjects: LCSH: Climatic changes–Health aspects. | Population–Health aspects.
Classification: LCC QC902.8 .S27 2022 | DDC 614.4–dc23
LC record available at https://lccn.loc.gov/2020025786

6048

Printed in the United States of America
24 23 22 21 20   10 9 8 7 6 5 4 3 2 1

This book is dedicated to my wonderful extended
loving family and especially my husband, my two
sons and my young grandchildren who will become
"30 somethings" around the year 2050. My hope
is that their world will be one where it is possible
to live a full and healthy life because adults alive
today will have shifted course and taken the
actions that are needed to secure the blessings of
a healthy life for the generations to come.

© BenGoode/Getty Images.

Brief Contents

© BenGoode/Getty Images.

Contents

What Is Happening

Prologue

Most Americans understand that climate change is real and are concerned about it.[1] However, many people still see climate change as a faraway threat, in both time and place, and as something that threatens the future of polar bears and glaciers but not necessarily people. The reality, however, is starkly different: climate change is already causing problems in communities in every region of our nation and around the world, and from a health perspective, *it's harmful.*[2-11]

Most people in the United States are not aware of the health harms of climate change. A recent survey showed that more than half of American adults have not considered how global warming might affect people's health, and few (32%) can name a specific way in which climate change is harming our health. Few are aware that some groups of people—including our children, the elderly, the sick, and the poor—are most likely to be harmed by climate change.[12,13] None of these survey findings are surprising. There has been limited public discussion of the health harms of climate change.

This is starting to change, with more people expecting to see health impacts within the next decade,[14] but there is a clear need to share our increasing understanding and concern about the health consequences of climate change with our entire population so everyone understands how it may affect them and their families and communities. All people living in the United States should understand the following evidence-based information:

1. **There is a scientific consensus about human-caused climate change.** The reality of human-caused climate change is no longer a matter of debate. Based on the evidence, more than 97% of climate scientists have concluded that human-caused climate change is happening. Many studies have proven this fact.[15]
2. **In communities across the nation, climate change is harming our health now.** Doctors are seeing the health of their patients being harmed. Public health professionals are seeing increasing rates of health problems associated with climate change in their communities. These harms include heat-related illness, worsening chronic illnesses, injuries and deaths from floods and dangerous weather events, infectious diseases spread by mosquitoes and ticks, illnesses from contaminated food and water, and mental health problems.[5]
3. **The health of any individual can be harmed by climate change, but some of us face greater risk than others.** Children, student athletes, pregnant women, the elderly, people with chronic illnesses and allergies, the poor, and those who have experienced environmental injustice are more likely to be harmed.[16]

4. **Unless we take action together, these harms to health are going to get much worse.** The sooner we act, the more harm we can prevent, and the more we can protect the health of people who live in our country and all other countries.[17]

5. **The most important action we can take to protect our health is to reduce heat-trapping pollution by reducing energy waste and accelerating the inevitable transition to clean, renewable energy.** Efficient buildings, neighborhoods that support not only transportation by automobile but also many other ways of getting around, and smart energy policies are all essential and achievable. Improved agricultural methods that sequester more carbon can also help. These transitions may be achieved with policy changes in building standards and codes, energy generation and utility regulation, transportation planning, and agricultural policy.

If successful, these transitions will not only limit climate change, but *prevent* additional damage to our climate. Accelerating a transition to clean energy has the added benefit of rapidly cleaning up our air and our water so that we can enjoy better health. Most people want clean air and water, and better health.[18-20] The other actions—addressing transportation, agriculture, energy efficiency, and the carbon footprint of buildings—would also help address the changes in our climate by reducing the output of "greenhouse" gases that are trapping heat in the atmosphere or sequestering carbon to prevent heat trapping. Carbon may be sequestered in plants or the soil to prevent heat trapping. The reduction of carbon pollution is called "*mitigation.*"

We also need to *protect* people from the potential health harms of climate change. When we intervene to protect people from climate change, we are engaging in "*adaptation.*" *Mitigation* and *adaptation* are the two main categories of climate change action that are addressed by policy. While these are often considered two separate activities, both are essential for addressing climate change.

Evidence and the Scientific Method

Overview

Understanding the science has allowed us to identify and understand how our world works. Understanding the environmental factors that produce our weather has made meteorology predictions far more accurate. A better understanding of the properties of materials has allowed us to live more comfortable lives protected from temperature extremes. Over time, new methodologies and insights have given us a detailed picture of the changes occurring now and over the last several thousand years in the physical, chemical, and biological realms that surround us. This work did not begin last year or the year before. The scientific facts of climate change were discovered as the result of rigorous work by many investigators documenting changes at multiple sites over time. The investigators found evidence of changes from many sources in nature that all pointed to one origin and one set of conditions. The sources of evidence confirming a warming of the earth in all of its diverse terrain come from every continent and from diverse geographies—the oceans, the land, and the atmosphere. This warming trend is causing widespread variations in the Earth's climate and its weather, which are affecting people's lives and ecosystems around the world. Those who grow food are facing changes in temperature and precipitation. Those who rely on fishing for their livelihood or to feed themselves are finding their food sources threatened from climate changes and other factors. Industries such as forestry and wine production that rely on cycles of nature are challenged by new conditions and threats.

The sources of evidence for climate change from around the world overlap and have been independently reviewed by scientists in the relevant disciplines and the national academies of science in developed countries around the world. While details may differ, the scientists have identified the same trends and reached the same basic conclusions about the origins and mechanisms. The Earth's atmosphere is changing as heat-trapping greenhouse gases have accumulated and continue to accumulate. The result is a warmer planet, which continues to become warmer and warmer.

Related changes are happening on the Earth's surfaces. Many species are threatened with extinction. Many of the Earth's principal ecosystems are endangered including coral reefs, rain forests, wetlands, and savannahs. The degrading environment is also a contributor to climate change. An example is the harvesting and/or burning of forests that could be absorbing the larger quantities of CO_2 that humans are emitting into the atmosphere. Another is the loss of wetlands that provide an overflow storage area for coastal deltas when rain increases and river flows rise leading to flooding. Environmental degradation contributes to climate change and removes some of the natural factors that protect communities from changing conditions.

Establishing the Evidence-Base

The work that is required to discover and understand these developing changes has been carried out by climate scientists committed to identifying and understanding the changes in the Earth's atmosphere, chemistry, and climate—and the impact on the living organisms that inhabit the world. As a result of this painstaking and rigorous work, climate scientists have generated many sources of overlapping evidence of global warming and its impact on our world over many years. These investigations and the explanations that accompany them are published.

The process of establishing scientific truth involves tremendous repetition. The input of multiple independent scientists repeating each other's work and then doing their own experiments to uncover underlying truths about the world is the basis of scientific progress. The process of repeating and reverifying outcomes and the peer review that assures accuracy and validity and precedes publication are key elements. This is how we establish scientific facts and how we understand how our world works.

Peer review is another essential part of this work. Before studies can be published, different climate scientists from those doing the initial work, and not their own associates, must review this work and offer their judgment about the relevance and accuracy of the evidence and the validity of the methods and conclusions. Different scientists perform the calculations to see for themselves whether they achieve the same results. When other scientists seek to understand whether the changes in nature can be explained by the same set of circumstances, they are checking the work of others.

Public health practitioners are often in the position of taking the evidence and translating it into public health practice. They may need to explain the evidence to the communities they serve. This leads to improved joint work by the community and public health organizations on the underlying problem challenging the community, and on developing or identifying the best and most appropriate solutions for that community.

Introduction

In Chapter 1 of this Primer, there is a review of the key evidence that scientists have uncovered, which enabled us to understand that the Earth's atmosphere is warming and that physical-chemical mechanisms are causing the planet to become warmer. This first section will explain what we mean by global warming and climate change. They are related terms that refer to the changes that we are experiencing on Earth. We will introduce several early scientists who laid the groundwork for understanding our atmosphere and the changes in it.

Chapter 2 addresses what climate change means for health. It will focus on seven health impacts of global warming. For each health impact, the book describes the nature or mechanism of that impact, what it means for health, and who is at greatest risk. Most of the content in the health chapter focuses on the United States, but there is some information in each section about global health impacts.

Chapter 3 helps you to communicate clearly about climate change and makes sure you understand what is most important to communicate to others about climate change and health.

Chapter 4 discusses policy solutions that are beneficial to health and especially relevant to the health sector. This is a topic that will expand greatly in the next few years.

References

1. Leiserowitz A, Maibach, E, Roser-Renouf C, et al. *Climate change in the American mind: November 2016.* Yale University and George Mason University. New Haven, CT: Yale Program on Climate Change Communication; 2017.
2. Sarfaty M, Kreslake JM, Casale TB, et al. Views of AAAAI members on climate change and health. *J Allergy Clin Immunol-In Pract.* Published online December 16, 2015.
3. Koh H. Communicating the health effects of climate change. *The JAMA Forum.* 2016;315(3): 239–240.
4. Sarfaty M, Bloodhart B, Ewart G, et al. American Thoracic Society member survey on climate change and health. *Ann Amer Thorac Soc.* 2015;12(2):274–278.
5. Crimmins A, Balbus J, Gamble JL, Beard CB, Bell JE, USGCRP. *The Impacts of Climate Change on Human Health in the United States: A Scientific Assessment.* Washington, D.C.: Global Change Research Program; 2016.
6. Wellbery C, Sarfaty M. The health hazards of air pollution—implications for your patients. *Am Fam Physician.* 2017;95(3):146–148.
7. Crowley RA, Health and Public Policy Committee of the American College of Physicians. Climate change and health: a position paper of the American College of Physicians. *Ann Intern Med.* 2016;164(9):608–610.
8. Ahdoot S, Pacheco SE, The Council on Environmental Health. Global climate change and children's health. *Pediatrics.* 2015;136(5):e1468–e1484.

9. Sarfaty M, Mitchell M, Bloodhart B, et al. A survey of African American physicians on the health effects of climate change. *Int J Environ Res Public Health*. 2014;11(12):12473–12485.

10. Policy of the American Medical Association, 2008 reaffirmed 2014; H-135.938; Global Climate Change and Human Health. https://searchpf.ama-assn.org/SearchML/policyFinderPages/search.action

11. Policy of the American Medical Association, 2016; H-135.923; AMA Advocacy for Environmental Sustainability and Climate. https://searchpf.ama-assn.org/SearchML/searchDetails.action?uri=%2FAMADoc%2FHOD-135.923.xml

12. Gamble JL, Balbus J, Berger M, et al. Populations of Concern. In: Crimmins A, Balbus J, eds. *The Impacts of Climate Change on Human Health in the United States: A Scientific Assessment.* Washington, D.C.: U.S. Global Change Research Program; 2016:247–286.

13. Yale Program on Climate Change Communication. Public perceptions of the health consequences of global warming. 2014. http://climatecommunication.yale.edu/publications/public-perceptions-of-the-health-consequences-of-global-warming/. Accessed February 19, 2017.

14. Leiserowitz A, Maibach E, Rosenthal S, Kotcher J, Bergquist P, Ballew M, Goldberg M, Gustafson A, Wang X. *Climate Change in the American Mind, April 2020.* Yale University and George Mason University. New Haven, CT: Yale Program on Climate Change Communication; 2020.

15. Cook J, Oreskes N, Doran PT, et al. Consensus on consensus: a synthesis of consensus estimates on human-caused global warming. *Environmental Research Letters*. 2016;11(4):1–7.

16. Gamble JL, Balbus J, Berger M, et al. *The Impacts of Climate Change on Human Health in the United States: A Scientific Assessment.* Washington, D.C.: U.S. Global Change Research Program; 2016.

17. American Association for the Advancement of Science. What we know: the risks, reality, and responses to climate change. https://whatweknow.aaas.org/. Accessed July 11, 2020.

18. Watts N, Adger W Neil, Agnolucci P, et al. Health and climate change: policy responses to protect public health. *The Lancet*. 2015;386(10006):P1861–1914. https://www.thelancet.com/journals/lancet/article/PIIS0140-6736(15)60854-6/fulltext

19. United States Environmental Protection Agency, Office of Atmospheric Programs. Climate Change in the United States: Benefits of Global Action, EPA 430-R-15-001.

20. Wang H, Horton R. Tackling climate change: the greatest opportunity for global health. *Lancet*. 2015;386:1798–1799.

Acknowledgments

There are many people to acknowledge for their wisdom and leadership on the science of climate change, the social science of communication, and for trailblazing on policy solutions to climate change. While admittedly a very brief exploration of these complex subject matters, this primer has benefited from the work of all of them. A few of those many individuals who conveyed their work to me directly include the following medical doctors and PhDs (listed alphabetically): Samantha Ahdoot, Fiona Armstrong, John Balbus, Ben Cash, Amy Collins, Howard Frumkin, Robert Gould, Edward Maibach, Mark Mitchell, Susan Pacheco, Cindy Parker, Jonathan Patz, Jerome Paulson, Wendy Ring, Linda Rudolph, Barbara Sattler, Nick Seaver, Jodi Sherman, Nick Watts, and Louis Ziska. I also wish to acknowledge and thank those who served on the United Nations Intergovernmental Panel on Climate Change and those who researched and wrote the 3rd and 4th U.S. National Climate Assessments.

For their assistance with this manuscript, I must acknowledge Edward Maibach for his extensive work on climate communication, Marybeth Montoro for her early work on the communication chapter, Kate Hoppe for her editing and for her work on the discussion questions, and Linda Rudolph for her work going back five years to develop a policy prescription targeted squarely on climate change and health.

Last but not least, I must acknowledge my husband who has encouraged me in my work on the climate crisis, put up with my long work hours, listened carefully, and frequently offered excellent advice.

About the Author

Dr. Mona Sarfaty is trained in family medicine and public health and has engaged in teaching, research, and advocacy for 40 years. As an academic faculty member with expertise in primary care, preventive services, and health policy, she has lectured at national and regional venues, including hospitals, health plans, professional societies, health departments, and government conferences. She is the founding director of the Medical Society Consortium on Climate and Health, a coalition of 29 medical societies and 55 affiliated public health organizations (and growing) with the mission of informing the public and policymakers about the health harms of climate change and the health benefits of climate solutions. Earlier in her career she worked as a Senior Health Policy Advisor for the U.S. Senate Health and Human Resources Committee (now Health, Education, Labor, and Pensions) where she planned hearings, wrote legislation, negotiated policy, and met with constituents. She collaborated with others to found the Foundation for the National Institutes of Health (NIH), the Community Oriented Primary Care Track at the George Washington University Milken Institute School of Public Health, as well as Project Access and the Primary Care Coalition of Montgomery County, Maryland, the College of Population Health, and the Diabetes Information and Support for Your Health programs at Thomas Jefferson University.

She is the author of widely circulated guides on cancer screening and many peer reviewed articles. She also authored two book chapters on climate change and health. She received her MPH from George Washington University, her MD from the State University of New York at Stony Brook, and her BA from Harvard University. She has been married for 38 years, and has two sons and four grandchildren.

CHAPTER 1

The Science Behind the Changing Climate

There is perhaps no better demonstration of the folly of human conceits than this distant image of our tiny world. To me, it underscores our responsibility to deal more kindly with one another, and to preserve and cherish the pale blue dot, the only home we've ever known.

Carl Edward Sagan, an American astronomer, author and science communicator, known widely as the co-writer of the television series "Cosmos: a personal voyage," the most widely watched television series in the history of public television.

KEY TERMS

Adaptation
Carbon Dioxide Concentration
Human Driver
Insect Vectors

Mitigation
Natural Driver
Ocean Acidification
Rising Seas

CHAPTER OBJECTIVES

1. Explain the chemistry that links the changes in the Earth's atmosphere to the Earth's temperature.
2. Describe the changes in the atmosphere of the Earth that are retaining more of the sun's warmth and how those changes have occurred.
3. Give three examples of measurement approaches that have allowed climate scientists to track changes in the composition of the Earth's atmosphere.
4. Give three examples of how the warming of the Earth affects human environments in ways that also impact health.

What Is Happening?

Fact: Ninety-seven percent of climate scientists understand that human-caused global warming (climate change) is happening.[1]
Fact: The National Academies of Science in every developed country agree that human-caused climate change is happening.[2]

Fact: The leading U.S. agencies that are responsible for space travel—the National Aeronautics and Space Administration (NASA)—and weather tracking around the globe—the National Oceanographic and Atmospheric Administration (NOAA)—are the sources of abundant evidence that defines the changes occurring in our atmosphere and our climate. These agencies are two of the most prominent science-based agencies in the world.[3]

Global Warming vs Climate Change

As you read this chapter, you will learn that the temperatures on the entire surface of the Earth, from the deepest places in the ocean to the highest mountain tops, are rising at an unprecedented rate. They are rising so quickly that humans, and the ecosystems that support human life are unable to adapt and they are impacted in ways that harm human health and well-being. A rise in the Earth's average temperature produces a cascading series of effects. The higher temperatures affect the balance of nature with the myriad parts of the larger ecosystem affected. This impacts human beings and all living things.

In other words, when the balance of nature is disturbed, this disturbance ripples through the smaller ecosystems that exist all over the Earth within the larger ecosystem. The impact is seen in the air, in the oceans, and on land (**Figure 1-1**). Each of these is the home of living things that are affected by the changes on Earth. The mechanisms that produce the disturbances in our ecosystems are not difficult to understand, but their interactions can be complex and some of the results may be unexpected.

"Global warming" is defined as the movement upward, over time, of the average temperatures in the atmosphere, on land, in the oceans, and on all various places on the Earth's surface. [4] As it was more frequently discussed, the term "global warming" was misconstrued. Some people mistakenly thought that it meant that *every day* in *every place* on Earth the temperature was warmer than it used to be. Since this was not the case and some days were consistently cold or even colder than average, people assumed that global temperatures must not be warming—and "global warming" was not a reality. Actually, an overall warming trend is consistent with days or even weeks or months that are below average temperature. Because "global warming" is the average of *all* of the temperature measurements on Earth, specific locations that are hotter or colder than average may not affect the *overall average*.

"Climate change" refers to the change in average weather patterns and climate systems over time.[5] The term "climate change" was coined to express the reality that the climate is different from what it was. Since it was believed that this term was easier for people to embrace than "global warming," it became the term of choice. "Climate change" and "global warming" are often used interchangeably, but have different technical meanings. Climate change means that many changes are happening, and people in different places may be seeing different changes. The meaning of climate change includes diverse and contrasting changes, such as more intense rainfall versus more intense drought or milder weather versus severe polar vortex freezes. "Global warming" and "climate change" both indicate a shift in our accurate understanding of Earth and they refer to the same large set of phenomena. This primer is about the impact on health because of those phenomena.

Similarly, the terms "weather" and "climate" are sometimes confused, although they refer to conditions with broadly different spatial scales and timescales."[5]

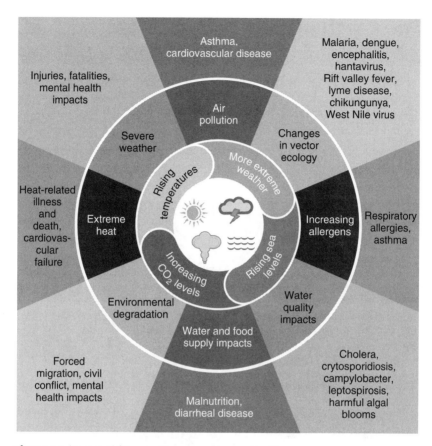

Figure 1-1 Impact of Climate Change on Human Health.

Reproduced from Centers for Disease Control and Prevention. *Climate Effects on Health.* https://www.cdc.gov/climateandhealth/effects/. Accessed September 14, 2018.

The weather changes from day to day: the temperature, the humidity, precipitation, etc. The climate describes those changes over a longer time period and larger area.

There is not one preferred term for the changes that are occurring. However, global warming (i.e., the increase in the average temperature of the atmosphere and on Earth's diverse terrains) is the underlying cause of the changing climate. Both terms refer to the significant changes in the temperature, weather, and distribution of land and water over the surface of the Earth. A group of mainstream medical, nursing, public health, and health professional societies and educational institutions began using the term "climate crisis" or "climate emergency" in 2019 as a more accurate term to describe the significant changes that are happening and how they threaten health (www.climatehealthaction.org).

This chapter describes the chemistry of the Earth's atmosphere, which has made Earth a planet that is warm enough while not being too hot to support human life. The chapter discusses 1) the changes that are occurring in the atmosphere, 2) the source of the changes, 3) how the changes are measured, and 4) the many changes that result from higher temperature on Earth and the environmental issues with health implications that stem from those higher temperatures.

The Earth and Its Atmosphere

Most likely, you remember some basic concepts about the Earth and its atmosphere from high school science. The earth revolves around the Sun. An atmosphere surrounds the Earth with a composition that has been carefully analyzed. All of the gases that comprise the Earth's atmosphere are in a band that is extremely thin and has many layers. In fact, the band is so thin compared with the Earth that it has been analogized to a sheet of saran wrap around a basketball.[6] See **Figure 1-2**. The images are of photographic silhouettes of the atmosphere surrounding the Earth. As thin as it is, this layer around the Earth holds all of the warmth that we experience unless we are using some other source of energy to keep us warm,

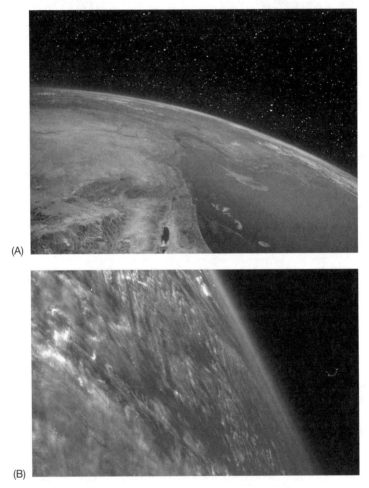

(A)

(B)

Figure 1-2 Without our atmosphere, the surface of the Earth, which comprises our immediate environment, would be much colder. We know exactly how cold on average that would be. It would be minus 18 degrees Fahrenheit (F) (minus 27 degrees Celsius, C), rather than the current average temperature of the earth, which is 59 degrees F (14 degrees C).

such as a fire or an electric heater. Without the atmosphere surrounding the Earth, more of the energy that the Earth receives from the sun would leak into space, and not create the warmth that we expect and experience. Without our atmosphere, the surface of the Earth—which comprises our immediate environment—would be much colder. Over time, we have come to understand exactly how cold, on average, that would be: minus 18°F (minus 27°C), as opposed to the current average temperature of the earth, which is 59°F (14°C).

How Does the Thin Atmosphere Surrounding the Earth Help Retain the Sun's Heat?

The composition of the thin atmosphere around the Earth makes life possible. Although there are five concentric layers of atmosphere around the Earth, those most important for human life are the two layers closest to the surface, the troposphere and the stratosphere. The layer at the surface that surrounds humans is the troposphere. The next layer out is the stratosphere.

The troposphere varies in height depending on the latitude of the Earth, stretching higher at the equator (at 65,000 feet, or >12 miles) than it does at the poles (at 25,000 feet, or just under 5 miles). It also changes by season, being shorter during winter. All weather on Earth, including the presence of clouds, occurs in the troposphere. The only exception to this is during the winter, when clouds also appear in the stratosphere, but this occurs only over the poles.

The next layer out, the stratosphere, stretches from the troposphere to 31 miles from the Earth's surface. Air is less dense and colder in the stratosphere. As compared with the stratosphere, the atmosphere's concentration of key components is greater in the troposphere, especially those that retain heat.

The atmosphere that surrounds the Earth consists of 78% nitrogen, 21% oxygen, and 0.9% argon (see **Table 1-1** below). There are other gases in even smaller trace amounts. These are: carbon dioxide, nitrous oxides, methane, and water vapor. The table below shows these components of the atmosphere and the relative percentages. Only some of these gases are responsible for the ability of the atmosphere to absorb infrared radiation (radiant heat) that comes from the Sun. In fact, it is only these "trace" elements in the Earth's atmosphere that can absorb the Sun's heat energy at all—and keep it within our atmosphere. Nitrogen and oxygen do not make the planet warm because they do not absorb the Sun's heat energy or radiate it back as heat.

On Earth, we see light from the Sun only when our side of the planet is facing the sun or when it is reflected off of the Moon. Our experience with heat is different. We even experience the heat from the Sun at night when we cannot see the Sun.

Table 1-1 Main Constituents of Earth's Atmosphere (Troposphere)

Gas	%
Nitrogen (N_2)	78.1
Oxygen (O_2)	20.1
Argon (Ar)	0.93
Carbon Dioxide (CO_2)	0.04 up to 0.100
Water (H_2O)	0.05 to 1.00

This is explained by the "trace" elements in the atmosphere. Humans' understanding of how those trace elements facilitate the warmth we experience on Earth is scientifically based.

During the 1820s, while thinking about how the Sun's heat could be retained to warm the Earth, the French physicist and mathematician Jean-Baptise Joseph Fourier (1768–1830) suggested the mechanism that turned out to be correct.[7] He believed that the atmosphere lets some of the Sun's radiation in but doesn't let all of it back out. The atmosphere traps some of the Sun's energy and that energy warms the surface of the planet.[8]

Forty years after Fourier completed his work, the Irish physicist John Tyndall identified which components of the atmosphere block the heat from escaping the Earth's atmosphere and are thus responsible for helping to retain heat at the Earth's surface. Tyndall studied the physical properties of air and infrared radiation (also referred to as radiant heat). He measured the gases' ability to absorb this radiant heat in the atmosphere. He discovered that water vapor, carbon dioxide, and ozone absorb heat radiation very well, while the two largest components of the atmosphere—nitrogen and oxygen—do not.[9] Before Tyndall, people believed that the entire atmosphere had a warming or "Greenhouse Effect." He was the first to offer evidence through his meticulous experiments and measurements that only certain atmospheric components have this heat-trapping warming effect.[10,11]

Figure 1-3 illustrates how the Sun's rays interact with the Earth. The light rays from the sun strike the Earth and the clouds; some bounce off and some are absorbed. The absorbed rays warm the planet, which then emits radiation of its own in the form of radiant heat (infrared). Carbon dioxide molecules, water vapor, methane, and nitrous oxide molecules in the atmosphere absorb this radiant (heat) energy and vibrate. Widespread vibration of these molecules in the atmosphere cause it to retain more of the Earth's radiant heat energy than it normally would. Each of these gases impact warming differently because of their physical chemical properties and because there are varied amounts in the atmosphere and they remain in the atmosphere for different lengths of time.

Nowadays, all atmospheric gases and their properties can be measured exactly. Because scientists now understand the composition and properties of the gases that comprise our atmosphere, they can identify and understand the temperature rise that has been occurring. Scientists can also predict anticipated effects. The changes in atmospheric components can be measured and analyzed and their impact can be predicted based on their physical properties.

The Observation of Global Warming

The Earth's average temperature has been rising at a rapid rate for the last 100 years. The geological and fossil records show that there have been changes in the Earth's temperature in the past; however, more recently, the *rapid pace of change* is unprecedented from when human life evolved. Of course, there is no *direct* way to measure the Earth's temperature 4.5 billion years ago. Nonetheless, study of the chemical structure and composition of rocks and other structures on the Earth's surface has enabled scientists to understand changes and developments in temperature, atmosphere, geology, flora, and fauna before human civilization. **Figure 1-4** shows the reconstructed changes in the atmosphere's composition and temperature from four hundred thousand years ago. The two graphs show that the temperatures on Earth track the carbon dioxide

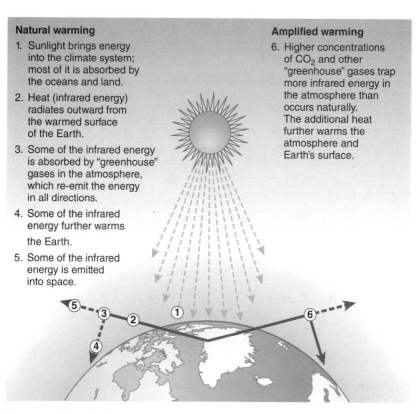

Natural warming
1. Sunlight brings energy into the climate system; most of it is absorbed by the oceans and land.
2. Heat (infrared energy) radiates outward from the warmed surface of the Earth.
3. Some of the infrared energy is absorbed by "greenhouse" gases in the atmosphere, which re-emit the energy in all directions.
4. Some of the infrared energy further warms the Earth.
5. Some of the infrared energy is emitted into space.

Amplified warming
6. Higher concentrations of CO_2 and other "greenhouse" gases trap more infrared energy in the atmosphere than occurs naturally. The additional heat further warms the atmosphere and Earth's surface.

Figure 1-3 How the Sun's Rays Interact with the Earth.

Reproduced from National Research Council. *Climate Change: Evidence, Impacts, and Choices: PDF Booklet*. Washington, DC: The National Academies Press; 2012. https://doi.org/10.17226/14673.

concentration of the atmosphere. Of course the carbon dioxide concentration is now well above the highest level shown here because these charts do not include the modern era. The zero line in the temperature chart represents the Earth's average temperature (59 degrees F or 14 degrees C). Although the temperature line on the graph is labeled "current temperature", the graph is out of date and the actual average temperature is higher than that line on the chart by one and half degrees Farenheit.

Figure 1-5 focuses on a more recent time period. It shows the measured temperatures that were warmer or cooler than usual from 1880 to the present, expressed as the difference from the average value for 1951–1980. The measured temperatures are from several different credible sources including NASA, NOAA, and U.S. authorities that predate these agencies, as well as a Japanese national authority. The graph's overall trajectory illustrates the rise in average temperatures over 140 years and shows how temperatures have changed since scientists started making direct measurements. The graph line shows predictable up and down variations since temperatures vary within a year as well as from year to year.

Between 1950 and 1980, the average temperatures were stable, going up and down but hovering around the same central number. This is labeled as 0 degrees on the graph. From around 1950 to around 1980, the cooler-than-average days were about as numerous as the warmer-than-average days. Before the stable years (1950–1980), average temperatures climbed and scientists debated about why temperatures were rising.

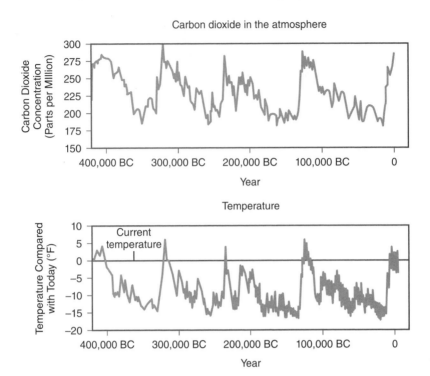

Figure 1-4 The Earth in the Past.

Reproduced from Environmental Protection Agency. A Student's Guide to Global Climate Change. https://archive.epa.gov/climatechange/kids/basics/past.html.

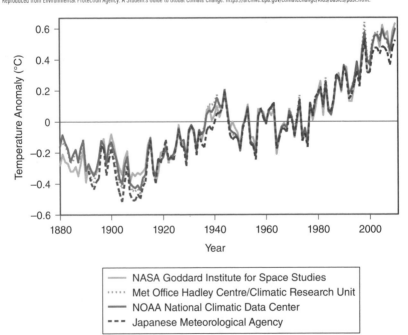

Figure 1-5 The Observation of Global Warming.

Reproduced from NASA Earth Observatory/Robert Simmon.

A century ago, Swedish scientist Svante Arrhenius recognized the increased output of carbon dioxide from burning fuel for power as a phenomenon that could warm the atmosphere and the planet.[12] In fact, during the 19th century, scientists already determined the basis for rising temperatures when they measured the increased carbon dioxide output that began with the industrial revolution.

However, some scientists believed that the oceans could absorb excess carbon dioxide and then the temperature would stabilize. To settle this debate, it became extremely important to capture accurate, standardized measurements of temperature and gas concentrations in the atmosphere. In 1958, there was worldwide agreement that the Mauna Loa Mountain Observatory in Hawaii would be the standard location for measuring the **carbon dioxide concentration** of the Earth's atmosphere. Since temperatures vary around the globe, another approach was needed. Temperature had to be measured at many different locations, at high and low points during the day, and averaged, to arrive at a daily number that represents Earth's average temperature.

How Do We Know that the Earth's Temperature Is Increasing?

Because temperatures differ all over the Earth, and the range of temperature is different in different locations and at different times of the year, a sound methodology is needed to determine the Earth's average temperature. Many different types of data have indicated that the world is warming. There are temperature sensors all over the Earth—and in the skies. Orbiting satellites, airplanes, ground sensors, and thermometers in diverse locations provide data that are measured over time, averaged, and compared. Some of these measurements come from NASA and some come from NOAA. The measurements are also compared with those made by other countries. **Figure 1-6** shows ten locations for measurements taken by multiple national authorities. The results have been consistent with each other and confirm rising temperatures.

Such measurements are essential to modern life. They allow prediction of global weather conditions on land and sea, and permit navigation of the skies and the oceans. The information that they yield allows us to live in the modern world.

How Do We Know that the Carbon Dioxide Level Has Risen Over the Past 2,000 Years?

You may wonder how we can know the level of carbon dioxide or other gases in the atmosphere during the 2000 years before scientists were able to measure them. There are several scientific approaches that have given us evidence to understand the carbon dioxide level in the atmosphere during the 2000 years before concurrent scientific measurement.

Carbon 14 dating is one example. It is used globally to learn the age of artifacts and archeological findings. Carbon 14 dating is effective because carbon 14 is formed continuously by interaction involving nitrogen in our atmosphere and neutrons from cosmic radiation. It integrates into the carbon dioxide (CO_2) in the atmosphere. If the CO_2 molecules are new, there will be more carbon 14 in them.

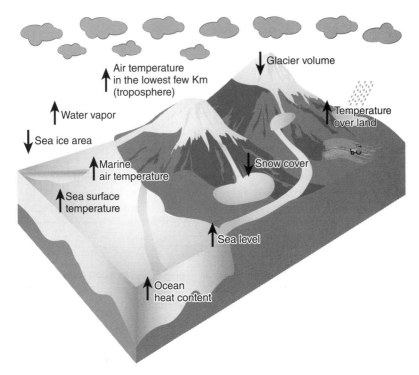

Figure 1-6 Ten Locations of Measurement to Confirm Rising Temperatures.

Data from IPCC. Climate Change 2014: Synthesis Report. In: Core Writing Team, R.K. Pachauri, and L.A. Meyer, eds. *Contribution of Working Groups I, II and III to the Fifth Assessment Report of the Intergovernmental Panel on Climate Change.* Geneva, Switzerland: IPCC; 2014:151.

If they are older, there will be less. Based on this technique, analysis of atmospheric carbon dioxide allows scientists to determine when the carbon dioxide in the atmosphere became part of the atmosphere.

The carbon 14 isotope of carbon breaks down over time and is replaced by other carbon isotopes. The half-life is about 5,700 years. If carbon 14 becomes part of a living organism, it is replenished. (It is no longer replenished when the plant or animal dies.) The ratio of carbon isotopes in objects enables scientists to calculate the age of objects from 500 to 50,000 years old. That technique, which is also used for dating archeological finds and animal remains, shows us that much of the carbon 14 in the carbon dioxide in the Earth's atmosphere has not degraded much and therefore, was placed there very recently.

The graph in **Figure 1-7** shows the increase in carbon dioxide in the atmosphere that occurred in two time periods: the mid-1700's until the late 1900's, and a shorter period of 28 years (1988–2014). The per year increase in the amount of carbon dioxide in the atmosphere was far greater in the more recent years. The difference is color coded with light gray and darker gray. During the nearly 250 years depicted in light gray, 737 gigatons of carbon dioxide entered the atmosphere; during the 28 years depicted in darker gray, 743 gigatons of carbon dioxide entered the atmosphere.

Another scientific approach to identifying the carbon concentration that existed in the atmosphere hundreds of thousands of years ago is chemical analysis of ice that is retrieved by drilling deep into the glaciers of Antarctica. Ice that freezes always contains air. The ingredients of the air in the ice can be measured

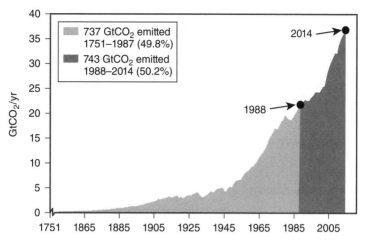

Figure 1-7 More than half of global CO_2 emissions (1751–2014) have been released since 1988.

Frumhoff PC, Heede R, Oreskes N. The climate responsibilities of industrial carbon producers. *Climatic Change*. 2015;132:157–171. https://doi.org/10.1007/s10584-015-1472-5.

by chemical analyses. Because Antarctica was frozen for so long, the ice beneath the surface has been unchanged for centuries. Scientists realized that if they could obtain pieces of that older ice, they could determine the contents of the atmosphere at earlier time periods when the ice had been formed. This insight and drilling deep into the ice for "ice cores," followed by analysis of this ancient ice, was a scientific breakthrough by Lonnie and Ellen Mosely Thompson, who received a top U.S. scientific award for their work, the Franklin Institute Medal. Analysis of the ice cores revealed that the CO_2 level in the air at the time when ice was formed was significantly lower. In fact, these cores give us information about the atmosphere dating back about 800,000 years.[13]

During the Industrial Revolution, in the mid 1700s, burning of fossil fuels to provide power to machines became common. From basic science, we learned that burning any organic material produces CO_2 and water as waste products. CO_2 in our atmosphere has been rising ever since we began burning fossil fuels for energy as a result of the Industrial Revolution.

Analysis of the air trapped in Antarctic ice that has remained frozen for over a thousand years, shows how much change has occurred in the amount of carbon dioxide in the air. Several ice core samples dating from the year 1000 to 1750 have given us a record of the concentration of carbon dioxide in the air at that time compared to now. They show that the concentration started to increase rapidly beginning around 1750.

Figure 1-8 shows data from ice cores. It illustrates a consistent CO_2 level mixed with the air from the ice dating from 1000 to 1750. The measurements are from several cores. The y-axis shows the concentration of CO_2 and the x-axis demonstrates the years of the ice core samples. In approximately 1750, the CO_2 level in the ice increased quickly over its previous level.

All of this evidence proves that the elevated carbon dioxide concentration in the atmosphere started at a time when human activity was adding carbon dioxide to the atmosphere and illustrates how the change in the concentration is the result of human activities.

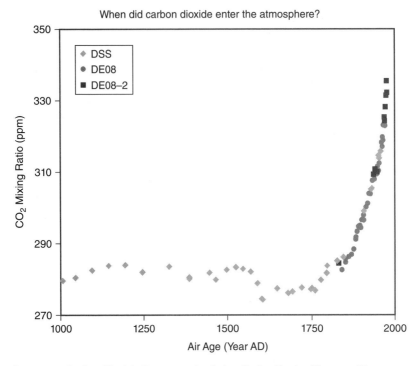

Figure 1-8 Carbon Dioxide Concentration in Ice Dating Back a Thousand Years.

Etheridge DM, et al. Historical CO₂ records from the Law Dome DE08, DE08-2, and DSS ice cores. In: *Trends: A Compendium of Data on Global Change.* Oak Ridge, TN: Carbon Dioxide Information Analysis Center, Oak Ridge National Laboratory, U.S. Department of Energy; 1998. https://cdiac.ess-dive.lbl.gov/trends/co2/lawdome. Accessed September 21, 2019.

Increasing Global Average Temperature and Carbon Dioxide Levels

Figure 1-9 illustrates how the Earth's temperature and the CO_2 atmospheric concentration changed during the same time period shown in the first graph. The x-axis shows the years. There are two y-axes that appear—on the right *and* on the left. The y-axis on the left shows how the actual average temperature changed from the late 1800s until now; the y-axis on the right shows how the CO_2 concentration of the atmosphere changed during the same time period. These measurements were recorded by scientists. They demonstrate the basic existing relationship in our atmosphere between the CO_2 concentration indicated by the line graph and the temperature indicated by the bar graph. In this graph, it can be seen that the temperature and the CO_2 concentration have risen in tandem.

Explanation from the U.S. National Climate Assessment: The average temperature of the world is already 1.5 degrees F (.8 degrees C) higher than it was between 1880 and 2012. Dark gray bars show temperatures above the long-term average, and light gray bars indicate temperatures below the long-term average. The black line shows atmospheric CO_2 concentration in parts per million (ppm). There are temperature fluctuations from one year to the next, but there is an obvious longer-term warming trend although not every year is warmer than the year before.

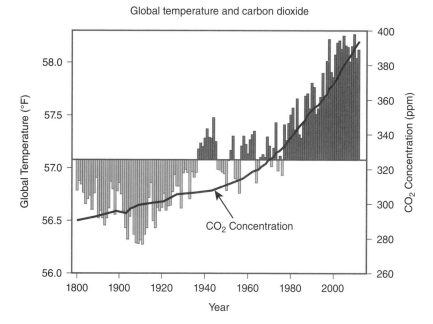

Figure 1-9 Global Average Temperature and Carbon Dioxide Levels.

The temperature fluctuations that occur from one year to the next are due to natural processes, such as the effects of El Niños, La Niñas, currents and volcanic eruptions.

Figure 1-9 demonstrates that the average worldwide temperature was 56.5 degrees F in the 1880s and rose to an average of 59 degrees F by 2018. During that same time period, the average carbon dioxide concentration rose from about 300 parts per million (ppm) to more than 400 ppm, and is now nearly 410 ppm.[14, 15] This is not a coincidence; the carbon dioxide concentration and temperature are linked. The graph shows this but does not explain the relationship; it shows only what has been measured.

The Role of Greenhouse Gases in the Changing Composition of Our Atmosphere

CO_2, nitrous oxide, and methane all play a significant role in warming our atmosphere because they absorb and re-emit heat. As their concentrations increase, they act as insulators that warm our planet. This is why they are called "greenhouse gases." **Figure 1-10** illustrates changes in the atmosphere on a larger time frame than the earlier graphs going back two thousand years (twenty centuries). The change in the rate of rise during recent years compared with the past is pronounced and significant. The x-axis shows the years. The y-axis shows levels of carbon dioxide and nitrous oxide on the left and methane on the right. Up to date measurements

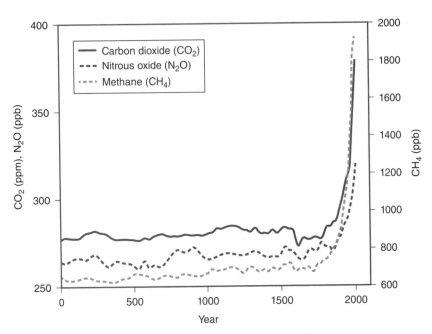

Figure 1-10 Concentrations of Greenhouse Gases from 0 to 2005.

Data from IPCC. *Climate Change 2013: The Physical Science Basis*; 2013. http://www.ipcc.ch/publications_and_data/ar4/wg1/en/fig/faq-2-1-figure-1-l.png.

are higher than they were in 2005. In 2020, Methane was 1876.3 ppb and nitrous oxide level was 334.04. (https://www.esrl.noaa.gov/gmd/ccgg/tre and https://www.n2olevels.org/).

CO_2 is a waste product from all combustion (burning) and is emitted whenever any fossil fuel, such as coal, oil, natural gas (methane), or wood is burned. CO_2 is the most common product of burning although methane and nitrous oxide are more potent greenhouse gases. Both methane and nitrous oxide exist in smaller quantities than CO_2 in our atmosphere but they have a greater warming potential.

Methane is the main ingredient in "natural gas," and escapes from natural gas drilling and flaring sites. It is also emitted by decomposing plant matter from various locations. Cattle release methane when they burp or pass gas after digesting their diet of plant matter. Decomposing plant matter is also found in landfills, swamps, rice paddies, and in the frozen ground of the Arctic called permafrost. When the permafrost melts because of global warming, more methane escapes into the atmosphere.

The overall quantity of methane emissions is less than CO_2 emissions but has a greater effect than CO_2. Each methane molecule has about 25 times the global warming potential of each CO_2 molecule, and it has the largest warming effect in the first two decades after its release.[16] Despite this substantial concern, natural gas is being used as a fuel more and more around the world.

Nitrous oxide has different origins. It is released from bacteria into the soil and from there can get into the atmosphere. In modern agriculture, emissions of nitrous oxide are increased because the additional nitrogen that is contained in fertilizers is transformed into nitrous oxide. One molecule of nitrous oxide has nearly 300 times the global warming potential of CO_2, and it remains in the atmosphere for 100 years. This is only one of the problems associated with use of fertilizers that contain nitrogen. Nitrogen also washes downstream with rainwater and contaminates waterways.

The rise in these three greenhouse gases, all of which absorb and reradiate the Sun's radiant (heat) energy, explains the increase in the average temperature of the Earth. It can be seen from the graph that the level of the three most notable gases: carbon dioxide, methane, and nitrous oxide, were relatively stable until the last couple of centuries when a steep and rapid rise became clear. Why might this be?

All three greenhouse gases are related to the growth of industrialization. The quantity of greenhouse gas that has accumulated in the atmosphere has come from multiple sources since the dawn of the Industrial Revolution. The carbon dioxide that entered the atmosphere in the United States in 2018 came from several major sources, specifically the energy burned to produce electricity (27%), to power transportation (28%) and industry (22%), to power residential and commercial uses (12%), and from agriculture fertilizer or the production of feed animals (10%). Worldwide, the sources are somewhat different. To reduce CO_2, methane, and nitrogen dioxide emissions, we need to greatly reduce the burning of fossil fuels to produce electricity and power transportation. We must also change agricultural practices. More efficient use of energy to provide heating and cooling for buildings, and for manufacturing of industrial products also has the capacity to reduce the use of fossil fuels.

There are other even more potent greenhouse gases used in smaller quantities that are produced synthetically, such as gases used for refrigerants, aerosols, and anesthetics. The use of aerosols and refrigerants has been addressed by international agreements. Hydrofluorocarbons are a notable example. Their source is the use of outdated refrigerants that remain in use. The Kigali Amendment to the Montreal Protocol was an international agreement on climate change that addressed this powerful greenhouse gas as part of the United Nations problem-solving work on the climate issue. Eliminating these emissions has been identified as one of the most easily achievable opportunities that the world has to mitigate the climate crisis.[17]

Natural Versus Human Causes of Atmospheric Warming

To explore every possibility as to why our atmosphere and the Earth have been warming, other explanations have been offered. These other explanations for the warming of the Earth and its atmosphere have been thoroughly explored. They include natural and human drivers.

Natural drivers. Natural drivers include natural variations in the Sun's energy, changes in the path of the Earth revolving around the sun that brings the Earth closer to the Sun at certain times based on its elliptical orbit, and natural events on Earth, such as volcanic eruptions that produce aerosols that block the Sun's rays. Several of these natural drivers are shown in **Figure 1-11**.

Scientists and mathematicians can measure the distance from the Earth to the Sun, changes in orbital revolutions around the Sun, volcanic emissions and other phenomena, and the estimated impact on the Earth's temperature. The estimates can be compared with the warming actually measured to determine if they are sufficient to explain the observed warming. However, if it were only those natural phenomena that warmed the Earth, the temperature would have changed in a way that is represented by the thicker light and dark gray lines on the graph in **Figure 1-12**.

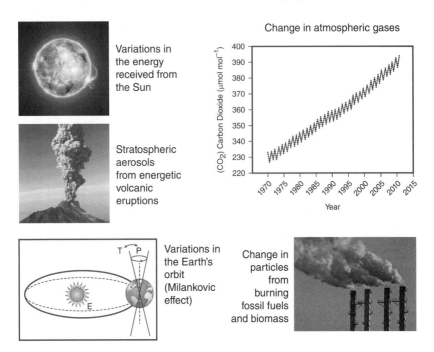

Variations in the energy received from the Sun

Stratospheric aerosols from energetic volcanic eruptions

Variations in the Earth's orbit (Milankovic effect)

Change in particles from burning fossil fuels and biomass

Figure 1-11 Natural Drivers of Climate Change Versus Human Drivers of Climate Change.

Top left corner: © Solarseven/Shutterstock; Middle left: © Aldivo Ahmad/Shutterstock.

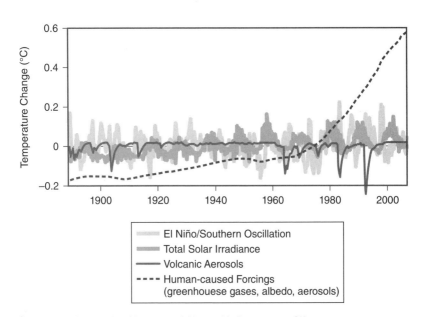

Figure 1-12 Separating Human and Natural Influences on Climate.

Reproduced from https://nca2014.globalchange.gov/report/our-changing-climate/observed-change#graphic-16679. Huber M, Knutti R. Anthropogenic and natural warming inferred from changes in Earth's energy balance. *Nat Geosci.* 2012;5:31–36.

The observed (measured) temperature is represented by the dotted line in Figure 1-12, not the gray lines.[18] No climate model that includes natural phenomena can explain the observed warming unless human factors are considered.

Human Drivers. The human drivers (Human-caused forcings) are the heat absorbing and heat reradiating gases that humans have added to the atmosphere as products of human activity. This is apparent in the graph. Without the human activity that has increased the heat reradiating gases (greenhouse gases) in the atmosphere, namely CO_2, methane, and nitrous oxide, warming would be far less. These gases have trapped the Sun's heat in the atmosphere that surrounds the Earth, preventing the heat from escaping into space and causing warming on every continent and every surface of the Earth.

The quantity of greenhouse gases that have accumulated in the atmosphere are from many sources of human activity since the beginning of the Industrial Revolution. Nearly all of the gases are from fossil fuels. As mentioned earlier, these U.S. sources of greenhouse gases are burning energy to produce electricity (27%), to provide power for transportation (28%), to power industry (22%) or other residential and commercial uses (12%), and for agriculture fertilizer or production of feed animals (10%). It should be noted that the emissions of the U.S. health care sector also account for approximately 10% of the total U.S. greenhouse gas emissions.[19]

In 1896, Svante Arrhenius suggested that burning fossil fuels such as coal could have an impact on the Earth's temperature because it would add CO_2 to the atmosphere. The idea was not well accepted at that time or a generation later. In the 1930s, measuring the temperatures in the United States proved that the country had warmed during the prior 50 years. At that time, this was commonly thought to be a phase of a natural cycle. The English engineer and inventor Guy Stewart Callendar published a paper suggesting that widespread warming was occurring because human activity was producing more CO_2.[20]

Scientists gather detailed records of how much coal, oil, and natural gas are burned each year. They have measured how much CO_2 is absorbed, on average, by the oceans and the land surface. These analyses show that approximately 45% of the CO_2 emitted by human activities remains in the atmosphere. Figure 1-7 shows the relative amounts of carbon dioxide emitted during two different time periods: 1750–1985 and 1985 to 2005. While CO_2 is the greenhouse gas in the atmosphere that traps the most heat due to its concentration and its longevity yet, it is not the only greenhouse gas. The other greenhouse gases emitted in the United States that have heat-trapping properties include methane gas (natural gas), oxides of nitrogen, and fluorinated gases. The fluorinated gases are 3% of emitted greenhouse gases and used as refrigerants. They are all more potent greenhouse gases than carbon dioxide.

Carbon Sinks

The specific greenhouse gases emitted in the United States and the source of those greenhouse gases are seen in the two pie graphs (See **Figure 1-13**).[6] Some gases in the first graph (Figure 1-13a) are a small portion of the pie but have a larger effect than other gases. Methane, for example, is more potent than carbon dioxide as a greenhouse gas and may increase faster than CO_2 as the earth warms because so much of it is currently trapped in the permafrost or frozen areas of the Earth. As these areas thaw, the methane concentration could rise even more.

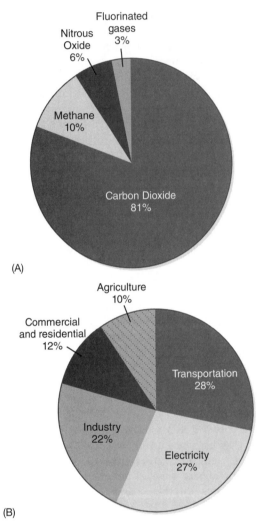

(A)

(B)

Figure 1-13 (A) Overview of Greenhouse Gas Emissions in 2018; **(B)** Sources of Greenhouse Gas Emissions in 2018.[21]

As the amount of CO_2 in the atmosphere rises, more is absorbed by natural systems. It is absorbed by plants, oceans, and soil. Plants on land and in the oceans use CO_2 for photosynthesis. When these die, the carbon is transferred to the soil or ocean. These natural systems that store carbon are referred to as "carbon sinks." They have the capacity to isolate some of the atmospheric carbon and have the potential to serve as part of the solution to rising CO_2 levels in the atmosphere.

While the CO_2 sinks help to lower the CO_2 in the atmosphere, the amount of CO_2 has surpassed the capacity of Earth's sinks to absorb it. In some cases, as with U.S. forests burning, the transition to cattle grazing with destruction of the Amazon jungle, or the destruction of the rain forests in Indonesia, major carbon sinks are being lost. This intensifies the problem of accumulating greenhouse gases. The result is the increased concentration of greenhouse gases in the atmosphere that are leading to a rise in the temperatures around the world.

The increase in greenhouse gases, loss of carbon sinks, and rising temperatures, are impacting the major components of our environment: the oceans and the land that serve as home to plants and animals, the air that supports plant and animal life, and the weather that causes changes in their day-to-day environment. These changes have implications for the growth and health of plants and animals, including human beings.

How Does the Warming of the Earth Affect the Environment?

We need to understand what is happening to our atmosphere and the Earth and how it got that way. But we also need to understand how these changes develop and interact and the implications of these changes for human health and well-being. The paragraphs below describe changes in the environment and how they can affect human beings. Chapter 2 focuses more specifically on the effects of climate change on human health.

Ocean Acidification. For years, some scientists thought that the oceans could absorb the excess CO_2 that people were putting into the atmosphere. In fact, a debate on this question was one of the factors that led the worldwide scientific community to agree on standardized approaches to measurement. The oceans have absorbed a significant amount of CO_2. But as this has occurred, the ocean's acidity has increased. In basic chemistry, CO_2 plus water leads to the formation of a weak acid (H^+) and an $^-HCO_3$ molecule. The acid dissolves some of the calcium in the shells of sea animals and can lead to their death. This affects clams, scallops, crabs, conch, nautilus, and so on. This undermines the shellfish supply that humans depend on for food. This process is also destroying coral reefs around the world. The destruction of coral is often referred to as a "bleaching process" because colorful coral turn white when they die. Major coral reefs around the world such as the Great Barrier Reef off the coast of Australia and major reefs in the Gulf of Mexico have lost two-thirds or more of the living coral. Since coral reefs are a fundamental part of the food chain of the ocean, the loss of coral reefs endanger the entire ocean ecosystem. This will adversely affect the fish as food sources for people around the world and contribute to world hunger problems.[22-26]

Rising Seas. Heat causes materials to expand and cold causes them to contract. This is also true for water. The Earth's changing temperature has been measured in the Earth's waters as well as in its atmosphere. The oceans have absorbed a good deal of the excess heat. This rising warmth has caused the oceans to expand and the sea level to rise.

Another factor that has occurred with rising temperatures is melting ice.[27] The ice sheet in the Arctic over the North Pole has diminished dramatically and is expected to disappear entirely in the not-too-distant future (see the graphic below).[28] When the ice melts, the darker oceanwater causes less reflection of the sun's heat and more absorption of the heat. This can speed up the warming of the oceans. Significant melting has also been occurring in Antarctica, although the warming in the Arctic and other northern latitudes has been greater than at lower

(A)

(B)

Figure 1-14 (A) Now you see it; **(B)** Now you don't.

Photo credits: Photographed by William O. Field on August 13, 1941 (top) and by Bruce F. Molnia on August 31, 2004 (bottom). From the Glacier Photograph Collection. Boulder, CO: National Snow and Ice Data Center/World Data Center for Glaciology.

latitudes. For example, the loss of ice has had a dramatic impact on ecosystems in Alaska. **Figure 1-14** is an example of the widespread glacial melting has occurred in Alaska. Native Alaskan villages are losing the frozen riverways that have served as avenues of transportation. Erosion at the edge of the sea has endangered coastal communities, and melting permafrost has softened the ground on which people have erected buildings leading to collapse in some cases.

The warming of the ocean and the addition of melting ice has caused the sea level to rise so much that the existence of some island nations is threatened. In the United States, the higher sea level has caused higher tides such that recurrent flooding at high tide is already occurring along the length of the coastlines. Along the coast of Virginia where there are important naval installations, the sea level rise has increased, and there is great concern about the military bases flooding. There is talk of moving them to new locations on higher ground. For example, the sea level is rising rapidly in the Hampton Roads area where the sea level has risen by

14 inches since 1950. Scientists from NOAA and NASA find that the speed of the rise is increasing. They are able to document this using satellites, floating buoys, and tidal gauges.[29] An island off of the coast of Virginia in the Chesapeake Bay called Tangier Island that is home to an historic fishing community is slowly disappearing. In Delaware, a popular coastal bicycle trail is no longer usable when the tide is in.

Some examples have received a lot of attention and have become very widely known. In Miami, Florida, main roads are being raised to a higher elevation to avoid being disrupted by this recurrent flooding. When "king tides" occur, based on the cycles of the Moon, this recurrent flooding is more pronounced. These changes are threatening coastal development and livelihoods, such as oyster fishing that is in estuaries and based on a balance between salt and fresh water. In some areas, salt-water intrusions are occurring in underground aquifers that provide drinking water or irrigation water for farming. Livelihoods from tourism are threatened as popular beaches erode with higher tides.

In some coastal areas, the land is sinking. This is called *subsidence*. This phenomenon explains the change in the sea level at the shore in some areas. Many of these impacts receive limited attention. People at various locations on the East Coast find themselves cut off from their homes at high tide and unable to get in or out. They may have to travel different routes on different roads. **Figure 1-15** depicts the changes in the ocean that are related to planetary warming; the figure includes actions that people can take to slow these changes.

Storms can compound these problems. When storms occur, storm surges can raise the water level further. Coastal flooding can enter or damage homes— extensively at times. Whether it is recurrent flooding or flooding resulting from a storm surge, water seeping into a building causes dampness and mold growth. Mold is a troubling allergen for some people and can cause attacks of asthma.

Extreme Weather. Our atmosphere naturally contains a small amount of water vapor or moisture. The amount of water vapor is generally reflected by the humidity. If the water vapor or moisture in the air increases to more than the air can hold, the moisture turns to raindrops instead of vapor and it rains. The moisture-carrying capacity of the air varies, but warm air can hold a good deal more moisture than cool air. As the atmosphere gets warmer, the quantity of water that the air holds is greater. For every degree of higher temperature, the air can hold 4–7% more moisture. When the moisture-carrying threshold is surpassed and it rains, the quantity of rain that falls can be much greater than usual because the moisture content of the air is greater than usual.

The graph shows the change in the number of heavy precipitation events over time from 1900 to 2000. See **Figure 1-16**. The increase is a consequence of the growing moisture content of the warmer atmosphere. Heavy precipitation events are defined as those in which the amount of rain that fell was greater than the 90th percentile for all rainfall events. The bars on the graph indicate the change in these events during the 20th century and show clearly that after 1950, there was a steady and dramatic rise in the number of extreme rainfall events.

Warmer seas radiate more heat and add more water vapor into the air adjacent to them than cooler seas. This added heat energy and differences in temperature can lead to high velocity winds that, along with increased rainfall from the added water vapor, contribute to hurricanes forming.

These phenomena occurred in 2017 when Hurricanes Harvey, Irma, and Maria reached higher velocities, and greater amounts of rain fell than with many previous hurricanes. In fact, Harvey produced the greatest amount of rainfall of any

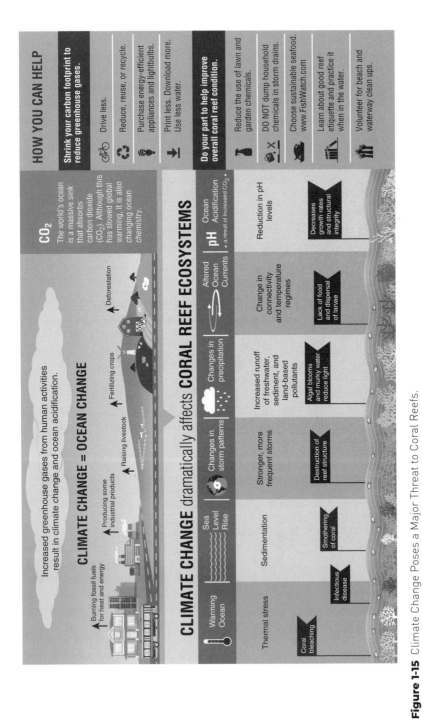

Figure 1-15 Climate Change Poses a Major Threat to Coral Reefs.

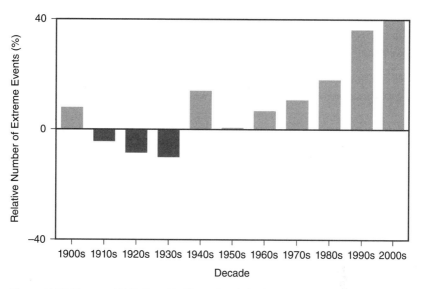

Figure 1-16 Observed U.S. Trend in Heavy Precipitation.

Reproduced from U.S. Global Change Research Program. *Climate Change Impacts in the United States: The Third National Climate Assessment.* https://www.globalchange.gov /browse/multimedia/observed-us-trend-heavy-precipitation.

storm in U.S. history.[30] Maria was one of the strongest and most rapidly intensifying hurricanes ever recorded, costing $90 billion in damages to the island of Puerto Rico.[31] Another factor associated with the increased precipitation levels is the slower moving nature of some storms, which has allowed them to stay in one place longer and rain more in a particular location. This slow-moving development is explained by changes in the jet stream's pattern as it moves across the United States. This is also thought to be linked to changing temperatures in the Arctic.[32]

With increases in the intensity of rainfall, there are also increased numbers of flooding events in many parts of the United States. These events do not occur because of sea level rise. These flood events are due to heavy rainfall. Flooding causes harm to people, property, and crops. There can be a direct loss of life from drowning as well as loss or damage to property. More people lose their lives due to flooding than from other extreme weather events, although in the United States as a whole, heat causes greater loss of life than rainfall and storms. Loss or contamination of crops with sewage or toxic substances also occurs during flooding and can impact the food supply.

Rising Temperature. The increase in the average temperature does not occur uniformly around the world. While the *average* change may be only one and a half degrees F warmer (slightly less than one degree C), individual locations may exceed average temperatures by quite a bit. New temperature records, both hotter and colder, are set more frequently as the *average* temperatures become warmer. Three of the warmest years in recorded history occurred between 2014 and 2017. Eighteen of the nineteen warmest years on record occurred from 2000 to 2019. Statistically speaking, when a normal curve moves one standard deviation from the mean toward the warmer end, the frequency of events at the extreme end of the curve becomes far more frequent. A one-in-40-year event may become a one-in-6-year event.

While temperatures are rising globally, local average changes in temperature may differ from each other based on several factors. The narrower height of the troposphere at the north and south poles is one factor. Another factor is the amount of water vapor in the air. There is less water vapor (moisture) in the air closer to the poles than farther south. Since water vapor moderates the rise in temperature, the rise in average temperature is 4 degrees F in Alaska but only 2 degrees F in other parts of the United States.[33]

As we will explain in the next chapter, increased heat can lead to direct harm to people, especially to those who spend more time outdoors or are more sensitive to heat consistent with their age (very young or very old), or have underlying health conditions or certain medication regimens. A combination of extreme heat and increased moisture in the air (humidity) can be especially dangerous since it becomes more difficult for sweat to evaporate. Sweat cools the body when it evaporates. Heat illness can occur with a variety of symptoms that may or may not be recognized. It can progress to heat stroke and death unless treated quickly. Being aware of the dangers of extreme heat and how to prevent it, including hydrating and cooling, is essential.

Drought and Fires. Not all precipitation changes will result in greater rain. Climate change is expected to increase the existing contrasts in moisture so that wetter areas become even wetter and drier areas become even drier. Higher temperatures can increase evaporation from the soil, plants, lakes, and streams. The water that runs off from rain and snowfall flows into rivers and creeks, and rising temperatures can lead to faster evaporation of this run off. Because of these factors, rising temperatures and evaporation can increase the chance of a drought. Drought makes fires more likely.

Scientists predicted that forests that were already fire-prone, such as the evergreen forests of the United States and the Canadian west, would become more fire-prone. It was predicted that every degree of warming would increase the area that burned by 2 to 4 times. The reality has been close to the prediction. Lately, fire seasons have come earlier, lasted longer, and destroyed more acres than any time in our recorded history.

An analysis in 2016 showed that the annual number of large fires has tripled since the 1970s, and the amount of land that those fires have burned is six times more than it was 4 decades ago. A report by the group Climate Central detailed 45 years of state-by-state wildfire trends on U.S. Forest Service lands. Climate Central provided the first state-based projections of fire risk moving forward in time.[34]

The average number of fires on Forest Service land has increased at least 10-fold since the 1970s in the Northern Rocky Mountain states of Wyoming, Idaho, and Montana. In Washington State, there are now five times as many large fires burning in a typical year than in the 1970s; in Oregon, there are nearly seven times as many.

The average number of large fires over 1,000 acres burning each year on U.S. Forest Service lands in western states is shown in **Figure 1-17**. The figure shows the number of large fires on the y-axis and the years on the x-axis. The rise in large fires represented by the bars over the time period from 1988 to 2008 is evident. There is also a black line across the graph that represents the change in average temperature during that same time period. In addition to higher temperatures, other factors, such as land use or tree diseases, play a role, but climate change accounts for close to half of the increase.[35] **Figure 1-18** depicts the proportion of increasing wildfires that may be attributed to climate change.

While fire is a direct threat to people and their homes, the smoke from a fire is filled with small particulates (particles) that can enter the smaller airways of the lungs and cause breathing difficulty and other issues. Since the smoke travels hundreds of miles, even people who are far from the fire may be affected. People who

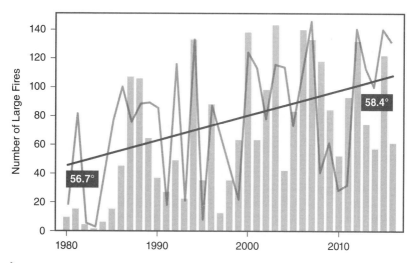

Figure 1-17 Hotter Years, More Fires.

have underlying asthma or chronic lung disease are especially vulnerable. About 15% of the population has some lung condition.

Air Quality, Pollen, Insect Vectors. Changes in warmth and the moistness of the air endangers health in other ways. Air quality is affected by chemical reactions; wildfires; and the density, range, and seasonality of pollen-producing plants that send pollen particles into the air in certain seasons and regions. The survival and reproduction of insect vector populations, or insect populations that carry infectious diseases, are also affected.

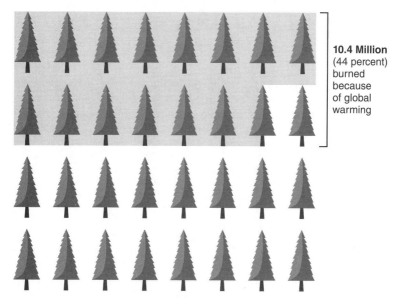

10.4 Million (44 percent) burned because of global warming

Figure 1-18 Proportion of wildfires attributed to climate change.

Air Quality. Air quality may be affected because heat and light interact with organic compounds and hydrocarbon fumes in the air and cause a chemical reaction that results in the production of ground-level ozone. We are *not* speaking here of the ozone layer in the upper atmosphere that protects us from the Sun's rays and was endangered some years ago and referred to as the "ozone hole." Here we are referring to the problem of "ground level ozone," or ozone gas in the air where people live, which has an entirely different impact. Ground-level ozone is a direct irritant to the bronchial airways of the lungs as well as other mucous membranes, such as the eyes. For those who have common lung conditions, such as asthma and/or chronic obstructive pulmonary disease, this irritation can cause breathing difficulties. If the concentration of ozone rises enough, even people without underlying lung conditions may be affected.

The air quality index offers a measure of the quality of the air. Two measures are often reported, ozone and particulates. Both can be problematic. Particulates will be greatly elevated as a result of fire. On the air quality index, when a "code red" for ozone is reached, children should not go outside for sports or school recess because of the health danger (see **Figure 1-19**). The dangers are great enough that, in some jurisdictions, public service alert messages discourage the local population from doing even minor activities that would add fumes to the air, such as driving their cars or using gasoline-powered lawn mowers.

Pollen. Global warming is lengthening the growing season, increasing the geographic range of many plants, and increasing the length of pollen seasons. The lengthening of pollen season is occurring in bands defined by latitude across the United States (see **Figure 1-20**). Pollen season is currently 10 days to 1 month longer than it was 20 years ago. The farther north in the United States, the greater the increase in the pollen season. Longer pollen seasons affect many different types of allergenic plants, from grass to trees, to wild ragweed.

Index values (conc. range)	Air quality descriptors	Cautionary statments for ozone
0–50 (0–59 ppb)	Good	No health impacts are expected when air quality is in this range.
51–100 (60–75 ppb)	Moderate	Unusually sensitive people should consider limiting prolonged outdoor exertion.
101–150 (76–95 ppb)	Unhealthy for sensitive groups	Active children and adults, and people with respiratory disease, such as asthma, should limit prolonged outdoor exertion.
151–200 (96–115 ppb)	Unhealthy	Active children and adults, and people with respiratory disease, such as asthma, should avoid prolonged outdoor exertion; everyone else, especially children, should limit prolonged outdoor exertion.
201–300 (116–374 ppb)	Very Unhealthy	Active children and adults, and people with respiratory disease, such as asthma, should avoid prolonged outdoor exertion; everyone else, especially children, should limit outdoor exertion.

Figure 1-19 Air Quality Index for Ozone.

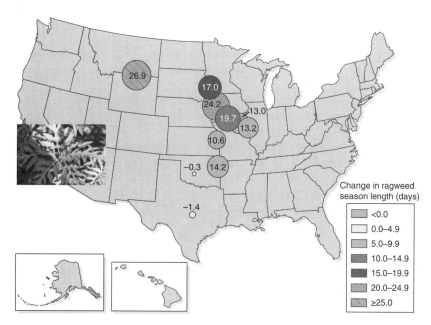

Figure 1-20 Ragweed pollen season length increased in central North America between 1995 and 2011 by as much as 11 to 27 days in parts of the United States and Canada, in response to rising temperatures. Increases in the length of this allergenic pollen season are connected with increases in the number of days before the first frost. The largest increases have been observed in northern cities.

Melillo JM, Richmond TC, Yohe GW, eds. *Climate Change Impacts in the United States: The Third National Climate Assessment*. Washington, DC: U.S. Global Change Research Program; 2014:842. http://dx.doi.org/10.7930/J0Z31WJ2.

Ragweed has been studied extensively. There is a ragweed belt that extends down the East Coast of the United States and stretches into the Midwest. This plant has been studied widely for its behavior in the air with different concentrations of carbon dioxide. Ragweed grown in air that is more carbon dioxide enriched produces more pollen. Furthermore, the pollen itself has been shown to be more allergenic.[36]

Physicians are reporting that their allergy patients now come in earlier in the year than they used to complaining of more symptoms than they used to.[37] The physicians have begun treating the patients earlier and with more medication than they did some years ago.

Another widespread, well-known plant that has been shown to be more allergenic when grown in air that has a high carbon dioxide concentration is poison ivy. The oil in the poison ivy plant that causes the rash, urushiol, is produced in greater quantities when grown in carbon dioxide-enriched air.[38] Thus, poison ivy is likely to produce more skin rash today than it did 10 or 20 years ago.

Opportunistic wild growth of plants that are not typically found in specific regions can also grow excessively as a result of changing conditions and ecosystems. This can transform landscapes in unwelcome ways to local residents. This phenomenon can be difficult to control. As ecosystems shift, the appearance of new animal species occurs as well. The spread of the deer population has contributed to the spread of tick-borne diseases. In this case, the lack of predators in the ecosystem is a contributing factor. Spread of disease carried by insect vectors is discussed below.

Vectors. Vectors are animals or insects that can spread disease. Ticks and mosquitos are two vectors that carry well-known diseases. Mosquitos carry West Nile virus, Zika virus, dengue fever, chikungunya, and malaria. Ticks carry Lyme disease, which is the most common and well-known tick-borne disease and happens to be spreading quickly.

The geographic range and reproduction of ticks and mosquitos are heavily impacted by changing heat and humidity and other factors.[39] Lyme disease is carried by deer ticks that become more numerous when winters are milder because young ticks do not die as quickly as they do when the winters are more severe. As the deer population multiplies and expands its habitat across many states because predators are scarce, the fact that they serve as a host for infected ticks, means that the tick population moves with them. Lyme disease is continuing to spread from the New England states to states that are further west, such as Michigan and Minnesota, north like Canada, and south, such as Virginia. Other rarer tick-borne infections have also appeared to a greater extent. These include ehrlichiosis, babesiosis, anaplasmosis, and other rarer diseases like Powassan's Disease.

Mosquito vectors that carry West Nile Virus, Zika virus, dengue fever, and chikungunya can be affected by temperature, humidity, and the population of people or animals that carry the disease. In 2016, an outbreak of Zika virus infection that spread from Brazil to Miami, Florida, caused numerous cases of serious neurologic disease affecting brain development in newborns and producing many cases of microcephaly with lifelong sequelae. The outbreak was contained but the danger was significant and threatened those populations who were vulnerable by virtue of geographic location.

West Nile Virus, another mosquito-borne infection, first appeared in the New York area in 1999 and then spread across the country, and continues to spread in southern states such as Texas and Louisiana. The spread of West Nile infections is consistent with bird migration routes, because the disease depends on birds as intermediate hosts for transmission. The continued spread of this infection in warm states 20 years after the first appearance in the United States is believed to be related to global warming.[40] Other mosquito born infections such as dengue fever and chikungunya depend on conditions that are ripe to breed the mosquito host. Both diseases have been reported in warmer southern states.[41,42]

Water Quality and Contamination. Heavy rains can easily disturb water supply and quality (and food supply as well). Too much rain accompanied by flooding creates the potential to overload storm water drainage systems, which can slow storm water runoff. The delayed water runoff leaves standing water, creating an excellent breeding ground since mosquitos deposit their larvae in water.

The storm water can cause flooding or it can mix with sewage or both, especially in the many places in the United States in which the sewage and storm water systems are not well separated. The local damage that flooding causes is tremendous. Buildings and livelihoods can be destroyed and water quality may be affected.

In farm regions, where fertilizers have been applied to crops, runoff of nitrogen in the fertilizer can provide nutrients for toxic algae growth in recreational lakes and navigable rivers downstream. Toxic algae can contaminate the water and make it undrinkable. The nitrates themselves can contaminate the drinking water.

Flooding in North Carolina after Hurricane Florence in 2018 was so heavy that it caused breeches of hog waste reservoirs and exposed local populations to become contaminated from hog feces.[43] During Hurricane Florence, people were also exposed to lakes of toxic coal ash.[44]

Crops can be impacted by such sources of contamination. Sewage mixes with storm water when the storm water and sewage runoff systems are combined and the combined system is stressed beyond its capacity. As a result, the food supply can be compromised by the flooding, the contamination, or both.

Drought, the Food Supply, and Food Insecurity. On the other side of the coin, air at a high temperature that is not carrying significant moisture can cause plants and soil to dry rapidly. This mechanism has caused devastating droughts in drought-prone areas of the United States and around the world. Such conditions threaten food and water supplies leading to devastating loss of livestock and crops. Endangered crops and livestock in turn leads to lower production, less availability, and higher food prices causing food insecurity. When the supply shrinks and prices rise, people who are hungry may no longer be able to adequately feed themselves and their families. They face hunger or starvation. These conditions affect the world because many countries buy grain from the United States or other countries.

Too little water presents other risks. Areas of the United States face severe shortages of drinking water. These shortages are expected to intensify. Infectious agents can concentrate in the lower water level in rivers or reservoirs, especially if water treatment is not optimal. This can contribute to the spread of contamination and cause waterborne outbreaks (see **Figure 1-21**).

Vulnerability to Climate Change and Regional Differences

Vulnerability to Climate Change. A valuable approach to understanding how climate change affects health is to consider specific exposure pathways and how they lead to human disease.[45] The concept of exposure pathways used in chemical risk assessment is helpful. The pathways illustrate the major avenues by which climate change impacts health. The exposure pathways can change over time and may be different in different locations. Climate change-related exposures can affect different people and different communities to a different extent. Furthermore, exposure to multiple threats can occur simultaneously, resulting in compounding or cascading health impacts. They may also accumulate over time, leading to longer-term changes in resistance and health.

Whether or not an individual person is exposed to a health threat or suffers adverse health outcomes depends on many different factors. Vulnerability includes three elements: exposure, susceptibility to harm, and the ability to adapt or to cope. All three of these elements can change over time and depend on the place and the situation. In a 2016 health report, the U.S. government defined the three elements of vulnerability as follows:[46]

- *Exposure* is contact between a person and one or more biological, psychosocial, chemical, or physical stressors, including stressors affected by climate change. Contact may occur in a single instance or repeatedly over time, and may occur in one location or over a wider geographic area.
- *Sensitivity* is the degree to which people or communities are affected, either adversely or beneficially, by a varying climate or other change.
- *Adaptive capacity* is the ability of communities, institutions, or people to adjust to potential hazards, to take advantage of opportunities, or to respond to

Figure 1-21 Climate Change and Health.

USGCRP. The Impacts of Climate Change on Human Health in the United States: A Scientific Assessment. In: Crimmins A, Balbus J, Gamble JL, Beard CB, Bell JE, Dodgen D, Eisen RJ, Fann N, Hawkins MD, Herring SC, Jantarasami L, Mills DM, Saha S, Sarofim MC, Trtanj J, and Ziska L, eds. Washington, DC: U.S. Global Change Research Program; 2016:312. http://dx.doi.org/10.7930/J0R49NQX.

consequences. A relevant term, *resilience*, is the ability to prepare and plan for, absorb, recover from, and more successfully adapt to adverse events.

These vulnerability components operate at multiple levels, from the individual and community to the country level, and affect all people to some degree. For an individual, these factors include human behavioral choices and the degree to which that person is vulnerable based on his or her level of exposure, sensitivity, and adaptive capacity. Vulnerability is also influenced by social determinants of health, including those that affect a person's ability to adapt, such as social capital and social cohesion (for example, the strength of interpersonal networks and social patterns in a community).

Regional Differences. Even though greenhouse gases in the atmosphere are distributed evenly, the impacts of climate change described above can vary considerably in different regions—in the country and the world. The way that climate change manifests—increased or decreased precipitation, drought, extreme heat, recurrent flooding, wildfires, etc.—will be heavily affected by location.[47] The effects of these changes will also differ based on the characteristics of individuals or populations. Some communities are in vulnerable locations, while others are less so. Some communities have fewer assets with which to adapt and protect themselves against adverse effects. Some have suffered environmental injustice and are vulnerable from the beginning. This is also true of individuals. Inequities leave people more vulnerable to begin with. Race and ethnicity continue to play a significant role in defining who is vulnerable. Some people have illnesses or conditions that make them more vulnerable. Age is a significant factor. Individual assets are important because they can make it possible for people to avoid being harmed or get the help that they need. These effects can interact in complicated ways and need to be understood to plan most effectively. The vulnerable people or communities have insights about these factors that others may lack.

Recap

As emphasized by the Centers for Disease Control and Prevention (CDC), complexity is a striking characteristic of climate change.[48] Many aspects of our world are changing at the same time due to the widespread effect of temperature in our environment and in our lives.

Oceans, land, air, weather, plants, animals, and people are all impacted by such changes. Feedback loops spread the impact in directions that are as complex as the ecosystems we inhabit. Despite this complexity, we must identify and understand the impact that they have on us and our life-sustaining systems to address them. We have seen how the Earth's atmosphere has determined its temperature and how the atmospheric changes affect the temperature. We have learned how the trace gases in the Earth's atmosphere are responsible for retaining the Sun's heat energy and how the increase in these, especially carbon dioxide but also methane and oxides of nitrogen, are trapping more of the Sun's radiant (heat) energy and causing all of the Earth's environments to become warmer. We have learned how some of these factors are measured and analyzed, even when the changes occurred thousands of years ago.

The environmental changes that stem from these changes to the Earth and its atmosphere have vast implications for human health. The environmental shifts include rising seas, extreme weather, heavy rains, floods, rising temperature, droughts, changes in air quality, pollen levels, insect vectors, and changes in water quality and the food supply. Different regions are confronted by different impacts depending on geographic location, weather characteristics, and the extent to which local areas have prepared for those impacts.

The preparations that are made to blunt the disruptive impact of these changes are jointly referred to as **Adaptation**. This is key to protecting populations from the extensive complex and interacting changes that are already seen across the country and across the world. **Mitigation** refers to efforts to reduce the underlying causes of climate change and thus prevent continued damage to the life sustaining systems that support human life. Mitigation actions include cutting the greenhouse gases that are entering the atmosphere and stopping the destruction of forests that protect the earth from the rising concentration of carbon dioxide. Adaptation is a significant aspect of what the United States and other countries must do to avoid the disasters that are already occurring because of climate change. The student is encouraged to consider the need and the work toward adaptation occurring in different regions of the United States depending on the major vulnerabilities they face. The Centers for Disease Control and Prevention support a program called Building Resistance Against Climate Effects or BRACE, which distributes funding to a third of the states to define vulnerabilities and build protective systems to make them resilient against these threats. Adaptation is not explored further in this text. Mitigation will be addressed in Chapter 4 on policy; resilience will also be discussed briefly in that chapter.

The next chapter will describe the health impacts of climate change and share some stories that will bring those impacts to life. The descriptions of health impacts have all been tested with readers to establish their clarity in describing the health effects of climate change.[49]

An appropriate response to these health harms and the conditions that produce them will require systems thinking, interprofessional cooperation, strategic planning, and best practices for communication. Communication will be the focus of the third chapter of this Primer. The fourth chapter will focus on policies that are needed to address climate change, most of which are also beneficial to human health.

WRAP-UP

Discussion Questions

1. Of the several approaches to measurement discussed in this chapter, which do you feel is easiest to understand and most revealing?
2. Which of the effects of global warming do you think are most likely to affect human health? Discuss these effects: ocean acidification, rising seas, extreme weather, rising temperature, more droughts and fires, poor air quality, and increases in disease vectors.
3. Some populations of people are more vulnerable to climate change than others. Which factors determine vulnerability?
4. What are the highest impact human drivers of global warming?

References

1. Cook J, Oreskes N, Doran PT, et al. Consensus on consensus: a synthesis of consensus estimates on human-caused global warming," *Environ Res Lett.* 2016;11(4):048002.
2. Gustin G. 22 National Science Academies urge government action on climate change. Inside Climate News. https://insideclimatenews.org/news/12032018/climate-change-solutions-national-science-academies-commonwealth-of-nations-paris-agreement. March 13, 2018. Accessed January 1, 2018.
3. National Oceanic and Atmospheric Administration: U.S. Department of Commerce. Nasa.gov; noaa.gov
4. National Geographic. Global warming. https://www.nationalgeographic.org/encyclopedia/global-warming/ (nd). Accessed September 21, 2019 and https://earthobservatory.nasa.gov/features/GlobalWarming. Accessed September 21, 2019.

5. Ibid.
6. National Aeronautics and Space Administration. The Atmosphere. https://www.grc.nasa.gov /www/k-12/airplane/atmosphere.html (nd). Accessed May 28, 2018.
7. Gunderman R, Gunderman D. On his 250th birthday, Joseph Fourier's math still makes a difference. The Conversation. March 21, 2018. https://theconversation.com/on-his-250th -birthday-joseph-fouriers-math-still-makes-a-difference-90906. Accessed July 25, 2020.
8. Ibid.
9. Graham S. John Tyndall (1820–1893). https://earthobservatory.nasa.gov/Features/Tyndall/ 1999. Accessed May 11, 2020.
10. Jackson R. *The Ascent of John Tyndall: Victorian Scientist, Mountaineer, and Public Intellectual.* United Kingdom: Oxford University Press; 2018.
11. Graham S. John Tyndall (1820-1893). October 8, 1999. https://earthobservatory.nasa.gov /features/Tyndall. Accessed July 25, 2020.
12. Enzler SM. History of the greenhouse effect and global warming. From Maslin M. *Global Warming, a Very Short Introduction.* United Kingdom: Oxford University Press; 2004. https:// www.lenntech.com/greenhouse-effect/global-warming-history.htm Accessed May 28, 2020.
13. CDIAC Carbon Dioxide Information Analysis Center. *800,000-year ice-core records of atmospheric carbon dioxide (CO$_2$).* 2012. https://cdiac.ess-dive.lbl.gov/trends/co2/ice_core _co2.html. Accessed September 21, 2019.
14. Lindsey R. Climate change: atmospheric carbon dioxide. https://www.climate.gov/news -features/understanding-climate/climate-change-atmospheric-carbon-dioxide. 2020. Accessed September 21, 2019.
15. Global Climate Change. Carbon dioxide. 2020. https://climate.nasa.gov/vital-signs/carbon -dioxide/. Accessed December 24, 2019.
16. Princeton University. A more potent greenhouse gas than carbon dioxide, methane emissions will leap as Earth warms. *ScienceDaily.* 2014. https://www.sciencedaily.com/releases/2014/03 /140327111724.htm. Accessed December 24, 2019.
17. Hawken P. Drawdown: the most comprehensive plan ever proposed to reverse global warming. 2017. New York, NY: Penguin Books.
18. Cash B. Center for Ocean, Land, Atmosphere. George Mason University. 2017.
19. Eckelman MJ, Sherman J. Environmental Impacts of the U.S. Health Care System and Effects on Public Health. *PLOS One.* 2016.
20. Weart SR. The discovery of global warming. The Center for the History of Physicians of the American Institute of Physics. 2017. https://history.aip.org/climate/summary.htm. Accessed May 28, 2018.
21. United States Environmental Protection Agency. Sources of greenhouse gas emissions. https:// www.epa.gov/ghgemissions/sources-greenhouse-gas-emissions. Accessed May 2020.
22. National Ocean Service. National Oceanic and Atmospheric Administration. U.S. Department of Commerce. How does climate change affect coral reefs? 2019. https://oceanservice.noaa.gov /facts/coralreef-climate.html
23. National Ocean Service. National Oceanic and Atmospheric Administration. U.S. Department of Commerce. How do coral reefs protect lives and property? https://oceanservice.noaa.gov /facts/coral_protect.html
24. National Ocean Service. National Oceanic and Atmospheric Administration. U.S. Department of Commerce. How do coral reefs benefit the economy? https://oceanservice.noaa.gov/facts /coral_economy.html
25. National Ocean Service. National Oceanic and Atmospheric Administration. U.S. Department of Commerce. How does overfishing threaten coral reefs? https://oceanservice.noaa.gov/facts /coral-overfishing.html
26. National Ocean Service. National Oceanic and Atmospheric Administration. U.S. Department of Commerce. What can I do to protect coral reefs? https://oceanservice.noaa.gov/facts /thingsyoucando.html
27. Center for Science Education. Climate and ice. https://scied.ucar.edu/longcontent/climate-and -ice. Accessed December 20, 2019.
28. Global Climate Change: Vital Signs of the Planet. Graphic: gramatic glacier melt. Now you see it, now you don't. *Global Climate Change: Vital Signs of the Planet.* 2020. https://climate.nasa.gov /climate_resources/4/graphic-dramatic-glacier-melt/. Accessed May 11, 2020.

29. Sea Level Rise.org. Overview: Virginia's sea level is rising. https://sealevelrise.org/states
    /virginia/?gclid=Cj0KCQiArozwBRDOARIsAHo2s7uKS_RgV3d-KfzowqM69qWId36U3-1kqC4
    Zy9hNDkjuuJF_3F8nfSYaArIvEALw_wcB. Accessed December 23, 2019.
30. Samenow J. Harvey is a 1,000-year flood event unprecedented in scale. *The Washington Post.* 2017.
    https://www.washingtonpost.com/news/capital-weather-gang/wp/2017/08/31/harvey-is-a-1000
    -year-flood-event-unprecedented-in-scale/?utm_term=.1099215d05bc. Accessed May 11, 2020.
31. Hersher R. Climate Change was the Engine that Powered Hurricane Maria's Devastating Rains.
    NPR WAMU April 17, 2019. https://www.npr.org/2019/04/17/714098828/climate-change
    -was-the-engine-that-powered-hurricane-marias-devastating-rains. Accessed July 25, 2020.
32. Nunez C. National Geographic. The jet stream, explained. 2019. https://www.nationalgeographic
    .com/environment/weather/reference/jet-stream/. Accessed February 15, 2020.
33. USGCRP, 2018: Impacts, Risks, and Adaptation in the United States: Fourth National Climate
    Assessment, Volume II [Reidmiller, Reidmiller DR, Avery CW, Easterling DR, Kunkel KE, Lewis
    KLM, Maycock TK, and Stewart BC (eds.)]. U.S. Global Change Research Program, Washington,
    DC, USA, 1515 pp.
34. Kenward A, Sanford T, Bronzan J. Western Wildfires: A Fiery Future. 2016. http://www
    .climatecentral.org/news/western-wildfires-climatechange-20475. Accessed September 13, 2018.
35. John T, Abatzoglou A, Park Williams. Climate change has added to western US forest fire.
    Proc Nat Acad Sci. 2016;20160701.
36. Ziska LH, Beggs PJ. Anthropogenic climate change and allergen exposure: The role of plant
    biology. *J Allergy Clin Immunol Pract.* 2012;129(1):27-32.
37. Sarfaty M, Kreslake JM, Casale TB, Maibach EW. Views of AAAAI members on climate change
    and health. *J Allergy Clin Immunol Pract.* 2015;4(2):333-335.
38. Carlberg R. Poison ivy potency? *Carnegie Museum of Natural History.* 2020. https://carnegiemnh
    .org/poison-ivy-potency/. Accessed December 23, 2019.
39. Centers for Disease Control and Prevention (CDC). Diseases carried by vectors. https://www
    .cdc.gov/climateandhealth/effects/vectors.htm. Accessed December 24, 2019.
40. Harrigan RJ, Thomassen HA, Buermann W, Smith TB. A continental risk assessment of West
    Nile virus under climate change. *Glob Change Biol.* 2014;20(8):2417-2425.
41. Centers for Disease Control and Prevention (CDC). Chikungunya virus in the United States.
    2014. https://www.cdc.gov/chikungunya/geo/united-states.html. Accessed July 15, 2019.
42. Centers for Disease Control and Prevention (CDC). Dengue. 2020. https://www.cdc.gov
    /dengue/. Accessed September 15, 2018.
43. Pierre-Lois K. Lagoons of pig waste are overflowing after Florence. Yes, that's as nasty as it sounds.
    2018. https://www.nytimes.com/2018/09/19/climate/florence-hog-farms.html. Accessed
    August 1, 2019.
44. Doran W, Murawski J. Duke Energy confirms coal ash spill in North Carolina. *News
    Observer.* 2018. https://www.newsobserver.com/news/local/article218718570.html. Accessed
    September 20, 2018.
45. Balbus J, Crimmins A, Gamble JL, et al, 2016: Ch. 1: Introduction: Climate Change and Human
    Health. The Impacts of Climate Change on Human Health in the United States: A Scientific
    Assessment. U.S. Global Change Research Program, Washington, DC, 25-42.
46. Ibid.
47. Balbus J, Crimmins A, Gamble JL, et al, 2016: Ch. 1: Introduction: Climate Change and
    Human Health. The Impacts of Climate Change on Human Health in the United States: A
    Scientific Assessment. U.S. Global Change Research Program, Washington, DC, 25-42.
48. Centers for Disease Control and Prevention (CDC). Climate change: communicating complexity.
    2009. https://blogs.cdc.gov/publichealthmatters/2009/11/climate-change-communicating
    -complexity/. Accessed September 21, 2019.
49. Kotcher J, Maibach E, Montoro M, Hassol SJ. How Americans respond to information about
    global warming's health impacts: evidence from a national survey experiment. *GeoHealth.*
    2018;2(9):262-275.

CHAPTER 2

Climate Change Is Harming Our Health

No matter where Americans live, climate change — through hurricanes, wildfires, heat, flooding, and the spread of disease — is harming our health, just as it is harming the health of people around the globe. We're all at risk and our leaders must lead on global warming.

Richard Carmona, M.D., was the
17th Surgeon General of The United States.
David Satcher, M.D., was the
16th Surgeon General of The United States.

KEY TERMS

Attention-Deficit Hyperactivity
  Disorder
Autism in Children

Decreased Cognitive Function
Neurodegenerative
Neurodevelopmental

CHAPTER OBJECTIVES

1. Describe what the population of the United States currently knows about how climate change (global warming) affects them.
2. Describe how climate change produces seven major health impacts.
3. Explain how and why health impacts vary in different regions of the United States.
4. Explain how the health impacts described here affect people around the world.

The harm to health and risks from climate change vary from region to region, but all regions of the country and all regions of the world are affected in some way. This text focuses on the impact in the United States, but the impact to human health that is described in this chapter occurs around the world. Low- and middle-income countries experience an impact to their health not generally seen in developed countries. Some impacts, such as the effects of extreme heat, extreme weather with flooding, and diseases caused by mosquitos, are far worse in low- and middle-income countries. Whether in the United States or in other countries around the world, health professionals must issue an alert and close the gap in notifying the public about this danger. The public trusts health professionals on this issue, and we must do everything we can to make sure that the public is aware.[1]

Most of the U.S. population takes climate change seriously.[2] However, many people still regard climate change as a threat in the future, in both time and place, and as something that threatens animals and plants but not humans. **The reality, however, is very different. Climate change is already causing problems in communities in every region of the United States, and from a health perspective, *it is harmful*.**[3-12] **Figure 2-1** shows the health risks people face across the U.S.

Although Americans increasingly believe that climate change is personally important to them and recognize that some people in their own community are affected by climate change, most are not aware of the detrimental harm that climate change causes to people's health.[13] A recent survey revealed that it has not occurred to most Americans how global warming might affect health, and few (<30%) can name a specific way in which climate change is harming our health. This is starting to change, but few people are aware that some groups of people—including

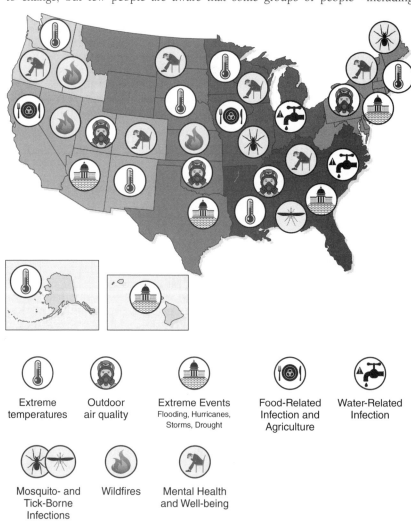

Figure 2-1 How Our Health Is Harmed by Climate Change.

Data from the Medical Society Consortium on Climate and Health, 2017.

children, pregnant women, the elderly, the chronically ill, and the poor and those experiencing environmental injustice—are most likely to be negatively affected by climate change.[14,15] People who have been adversely affected by environmental risks in the past are likely to be hurt the most and the quickest. The results of the survey are not surprising. In-depth reporting or public comment regarding the harm to health from climate change has been limited. But most survey respondents believe that doctors and nurses are trusted sources for information about health topics and for the topic of climate change in particular.[16-19] Health professionals, especially nurses, doctors, and pharmacists, are more trusted than other professionals in any other industry,[20] especially on health issues.

Because the substantial health impact of climate change will increase and worsen unless dramatic changes occur in fossil fuel use, deforestation, agricultural practices, and in adaptive capacity across the United States and around the world, it is essential that health professionals become more educated and share their concerns with the public. As discussed in Chapter 1, both mitigation and adaptation are necessary. Mitigation addresses and eliminates the direct causes of climate change; adaptation minimizes the adverse effects through preventive and protective measures.

Climate Change Is Harming the Health of People Across the United States

The Health Impact of Climate Change

Some of the physical changes in our climate—such as more heat waves, extreme weather, flooding, and air pollution—are causing direct **harm to health**. The big physical changes to our climate are also contributing to the spread of disease and threats to nutrition and mental health.

Although the entire world experiences these effects and we mention global impacts, in the chapters ahead we mainly summarize what is happening to the climate in the United States, describe how these changes are harming our health, and identify the specific U.S. population that is most likely to experience the greatest negative impact. Several of the harms described in this chapter are accompanied by stories of physicians who have directly experienced the impact in their own lives, medical practices, or communities.

Extreme Heat

What Is Happening?

Climate change is causing many more extremely hot days; greater humidity; and longer, hotter, and more frequent heat waves.

How Does that Harm Our Health?

Extreme heat can lead to heat-related illness and death from heat stroke and dehydration. It can also cause some chronic heart and lung diseases to worsen. Approximately 15% of the population has a chronic heart or lung disease. Individuals with these or other chronic, underlying health conditions can be at greater risk during a heat wave. A healthy person whose body is functioning well will sweat and their heart rate will increase when exposed to heat. Increased heart rate will bring the blood to the skin where cooling takes place at a faster rate. The sweat on the skin evaporates and cools the body, keeping the body temperature in a safe range. If sweating is decreased because of age, health conditions, or medications, the danger of heat will increase. Increased heart rate is compensatory but may also present a problem. Although the increased rate is part of a normal physical response to heat, it can present a danger if there is an underlying heart condition. In this case, heat can stress the heart and circulatory system, which keeps the blood moving. When the humidity is increased, the sweat may not evaporate as well, which decreases the cooling benefit that should occur. The heat index is a measure of the combined effect of the heat and the humidity.

When the heat index is elevated, the risk of higher temperatures as measured by the thermometer is actually greater than it would be otherwise. Those at increased risk should be listening, watching, or looking for reports on the heat index. These are typically carried in the newspaper, television news, or in meteorology reports. When the heat index is in the danger zone, individual activities should be curtailed and greater attention should focus on keeping cool and staying hydrated. Vulnerable people include children who spend more time playing or doing sports outdoors, pregnant women whose pregnancies can be adversely impacted by heat, outdoor workers, the elderly who sweat less, and the homeless.

When the body's systems are unable to compensate or adjust to keep the body temperature in a safe range, people may feel nauseous, shaky, dizzy, or light-headed and they may experience muscle cramps. This is called "heat illness" or "heat exhaustion." If left untreated, heat illness can progress to delirium or coma, which are referred to as "heat stroke"—and to mortality.

Who Is Being Harmed?

Everyone is susceptible to extreme heat, but some people face a greater risk than others. For example, outdoor workers, students who practice sports outdoors, city dwellers, and people who lack air conditioning (or who lose it during an extended power outage) face a greater risk because they are more exposed to extreme heat. People with chronic conditions such as cardiovascular and respiratory conditions may be especially vulnerable to extreme heat.

Heat can impede the functioning of damaged or inflamed lungs.[21] Poor air quality containing particulates from wildfire smoke, or a higher concentration of ozone (formed when light plus heat act on fumes in the air), causes those with lung conditions to be vulnerable. An otherwise stable chronic lung condition may start to destabilize under those circumstances.

Young children, older adults, and people taking certain medications are also more vulnerable because they are unable to regulate their body temperature easily. Children take more breaths and inhale more pollutants per minute. Because of that, they are more affected by climate-associated health impacts—and they have the most at stake over time as temperatures continue to rise. Also, pregnant women are vulnerable because air pollutants or extreme heat can lead to premature birth.[22,23]

Isaac's Story

By Dr. Samantha Ahdoot, Lead Author, American Academy of Pediatrics'
Policy on Climate Change; Pediatric Associates of Alexandria. Medical Society
Consortium on Climate and Health.

My nine-year-old son Isaac was attending his last day of band camp when I received a call from the emergency room. He had collapsed in the heat and was rushed to the emergency room. When my husband arrived at the hospital, Isaac was on a gurney with an IV in his arm, recovering under the watchful eyes of nurses and doctors. It was a terrifying experience for him.

That day was part of a record-setting heatwave in Washington, D.C., one of several days that summer when the heat index reached over 120 degrees.

As a pediatrician, I know that Isaac is not alone in his vulnerability to the heat. Emergency room visits for heat illnesses increased by 133% across the country between 1997 and 2006. Almost half of these patients were children and adolescents.

In August of 2010, another record hot summer, a colleague treated Logan, a young football player, in Arkansas. He showed initial signs of heat illness— weakness and fatigue—during practice in his un-air-conditioned gym, but he wasn't treated right away. He subsequently developed heat stroke, kidney failure, and pulmonary edema.

Fortunately, kidney dialysis saved him, but it was a close call.

Every summer, I see the impact of increasing temperature and heat waves on children like Logan and warn parents of the dangers of increasing heat waves.

I believe it's imperative that pediatricians on the frontlines of this urgent problem speak out for children on issues that will harm the health and prosperity of our youngest generations.

Fact: Heat illnesses are a leading cause of death and disability in young athletes. Approximately 9,000 high school athletes are treated for heat-related illnesses each year.[24]

Fact: *Young* men make up one-third of all heat-related emergency room visits in the United States.[25]

Fact: Football players may feel the most heat. They are 11 times more likely to suffer exertional heat illnesses than players of all other high school sports combined.[26] This fact is illustrated with **Figure 2-2**.

Figure 2-2 Youth football players outside in the heat.
© Brad Sauter/Shutterstock.

Global Impacts of Heat

From 2000 to 2016, the number of vulnerable people exposed to heatwaves around the world increased by approximately 125 million, with a record 175 million more people exposed to heatwaves in 2015.[27] The nature of the impact to health from too much heat exposure around the world is the same as in the United States. Heat illness, heat exhaustion, and heat stroke are all part of a continuum of symptoms, which can result in death. As indicated above, those people at greatest risk are infants, children, pregnant women, the elderly, those without air conditioning where it is otherwise standard (generally poor or low income people), and those who work outdoors.

The increase in chronic kidney disease of unknown origin occurs to a greater extent in the hotter regions of the world, although it is also increasing in the United States.[28] Experts have been monitoring this development, which suggests that climate change has become a major contributing factor to the spread of serious kidney disease.[29] The increase in incidence of kidney failure was first noted during the 1990s among those who work on fruit plantations in Central America where they are exposed to high temperatures for long time periods.

The average rise in worldwide temperature was 1.6 degrees F (.9 degrees C) between 2000 and 2016. Nineteen of the hottest years on record occurred between the years 2000 and 2019. However, *exposure* to increased temperatures is rising faster than the temperature increase. According to the Lancet Count Down report 2017, the average number of people exposed to warming temperatures has been rising faster more recently. This is because of the rapid rise in population density in India, China, and Sub-Saharan Africa. The increase in temperature in more populated areas is also greater than the overall average increase.

In many parts of the world, people are not protected against hot weather. During the heat wave in the summer of 2003, in Europe, extreme heat became deadly when more than 30,000 people died. Even economically developed nations like Germany and France did not have air conditioning in their homes because it was unnecessary in their milder climate up to that point. Heatwaves such as that in 2003 demonstrate that the mild climate is changing. In general, however, people in middle and low-income countries are more exposed. More than 2,300 people died in India in one month in June of 2015 during a heat wave.[30] Temperatures in certain areas of Pakistan are now routinely over 120 degrees F.[31]

Extreme Weather

What Is Happening?

Climate change is causing increases in the frequency and severity of some extreme weather events such as heavy downpours, floods, droughts, and major storms.

How Does that Harm Our Health?

Extreme weather events can cause injury, displacement, and death, and they can knock out power and phone lines, damage or destroy homes, and reduce the availability of safe food and water. Stomach and intestinal illnesses often increase following extreme weather and associated power outages. These extreme weather conditions can damage roads and bridges, impeding access to medical care and separating people from their medicines. Health facilities may also lose power or become unable to function because of flooding or other damage.

Who Is Being Harmed?

Everyone is vulnerable and can be harmed by extreme weather events, but emergency evacuations pose extra health risks to children, older adults, the poor, and those with disabilities (moreso if they are unable to access elevators and evacuation routes). People who are institutionalized can face exceptional barriers. After Hurricane Irma in 2017, a power outage in Florida caused the death of 12 nursing home residents who could not be easily evacuated.[32]

The powerful hurricanes of 2017, Harvey, Irma, and Maria, received a great deal of attention for powerful wind speeds, unprecedented rainfall, and a tendency to linger in one place, causing heightened damage. That year, natural disasters, including these three hurricanes, contributed to an overall cost to the United States of more than $300 billion. Prior storms with such destructive power occurred five years earlier in 2012 with Superstorm Sandy in the New York, New Jersey, and Connecticut (tri-state area); and seven years before that during Hurricane Katrina in New Orleans and the Gulf Coast in 2005—a storm that caused more than $125 billion dollars' worth of damage. The frequency of such devastating storms has increased as the climate has gotten warmer. In the Atlantic Ocean, north of the equator, powerful Category 5 hurricanes were coming on average about one time every 3 years (34 in 98 years). But the years 2016 through 2019 produced the longest sequence of consecutive years that all featured at least one Category 5 hurricane each. Five storms reached that category in the years 2016 to 2019—Matthew, Irma, Maria, Michael, and Dorian.[33]

Extreme rain storms defined as those greater than the 95th percentile have been steadily increasing all over the United States during the last sixty years. The next story highlights an extreme rainstorm in Louisiana referred to as a "thousand-year flood" because there was a less than 0.1% chance of a storm like this happening. This was the eighth, five hundred-year rainstorm event in the United States in that year.[34] **Figures 2-3** and **2-4** illustrate the rescue efforts that are made during such flooding.

Figure 2-3 Rescues such as the one shown here may be needed when stronger storms lead to flooding.

© Marcus Yam/Los Angeles Times/Getty Images.

Figure 2-4 Extreme storms and torrential rains are increasing across the country presenting greater risk of flooding.

It Rained in Sheets for Days

By Dr. Claude Tellis, Vice Chairman, Commission on Environmental Health, National Medical Association (NMA). Retired Pulmonologist. Medical Society Consortium on Climate and Health.

The damage from the deadly Louisiana flood of 2016, which struck my hometown of Baton Rouge and surrounding parishes in August, was still visible long after the rains stopped. Months later, homes were still gutted, and refrigerators, washing machines, and armchairs remained piled high on roadsides.

When Baton Rouge was hit with this "thousand-year flood"—a storm that has only a one-tenth of one percent chance of occurring in any given year—it rained in sheets for days. In the worst natural disaster since Hurricane Sandy, 13 people died, the Coast Guard rescued 30,000 people, and 10,000 people ended up in shelters. Some 180,000 homes and buildings were damaged.[35]

The storm also unleashed a health crisis for survivors. Some fleeing their flooding homes lost their medications for hypertension, diabetes, and heart problems. Others reported stress, depression, and anxiety in the weeks and months that followed. And long after the storm passed, some teachers reported children who felt so anxious and afraid when it rained that they needed counseling.

In the weeks following the storm, pools of standing water provided the perfect breeding ground for mold and mosquitoes. This was a problem because mosquito

Figure 2-5 Dr. Claude Tellis, Retired Pulmonologist and Critical Care Physician, Louisiana.

Courtesy of Dr. Claude Tellis.

borne illnesses like West Nile Virus were already found in the area. In some homes, mold from water damage still makes the air less healthy. We continue to struggle with the aftermath of this historic flood.

I believe our lives in Louisiana may never be the same because we will see continued suffering from the physical and mental damage of extreme weather—which is happening more often and with greater strength due to climate change. See a picture of Dr. Claude Tellis, in **Figure 2-5**. **Figure 2-6** shows how much of the state was flooded.

Global Health Impacts of Extreme Weather

A long-term upward trend has emerged in the number of flood-related and storm-related disasters in Africa, Asia, and the Americas since 1990.[36] This increase has reached statistical significance. Due to basic (simple) or absent systems that protect the infrastructure, the economic cost in low-income countries was almost double the cost in high-income countries, although the cost in higher-income countries was also substantial.[37] An example of the high cost in higher income countries is that the United States sustained $300 billion in weather disaster related losses in 2017 alone.[38] When homes and crop lands are destroyed, people lose their homes and livelihoods. An environmental crisis can easily turn into an economic and humanitarian disaster.

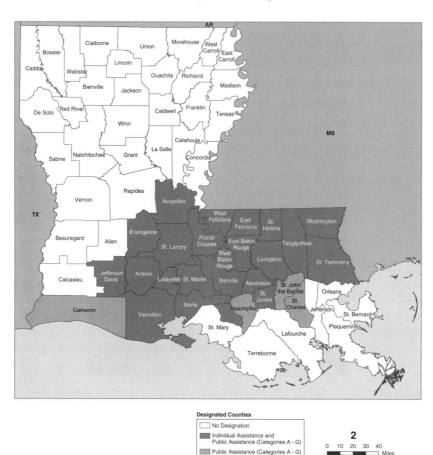

Figure 2-6 Twenty-two Louisiana parishes were designated as federal disaster areas by FEMA in the aftermath of the floods of 2016.

https://www.fema.gov/disaster/4277.

A few disastrous storms of the last few years have brought to light just how destructive strengthened hurricanes and typhoons (hurricanes of the Pacific Ocean) can be. Hurricane Dorian in 2019 with sustained winds of 185 mph completely decimated one of the islands in the Bahama Island chain.[39] It was the second strongest Atlantic hurricane in modern meteorological history. Typhoon Haiyan did the same type of damage in the Philippines in 2013 with maximum winds of up to 195 mph and gusts of up to 235 mph. It was one of the strongest storms ever recorded. The destruction and human cost of these disasters was staggering.

Interestingly, however, there is no documented increase in mortality that has accompanied these more numerous events and accompanying economic damage.[40] It is not clear why this is so.[41] It is possible that international and national emergency systems have protected the affected populations. Improved reporting systems may be playing a role in the count of events; the absence of mortality impact could be because of changes in reporting systems. These data may shift in the future such that increased events are specifically associated with a rise in seriously affected populations and mortality.

Air Pollution

What Is Happening?

Climate change reduces air quality for several reasons. Climate change is increasing smog, wildfires, and pollen production. As discussed in Chapter 1 in the section "Air Quality, Pollen, Insect Vectors," air quality is affected through chemical reactions, wildfires, and the density, range, and seasonality of pollen-producing plants that send pollen particles into the air in certain seasons and regions. Air quality may be affected because heat and light interact with organic compounds and hydrocarbon fumes in the air and cause a chemical reaction that results in the production of ground-level ozone. Ground-level ozone directly irritates the bronchial airways of the lungs as well as other mucous membranes including the eyes. Besides ozone, air particulates are the other major component of air pollution. Smoke from fires is filled with particulates. Pollen season is currently 10 days to a month longer than it was twenty years ago. This can also add substantial particulates to the air. The farther north in the United States, the greater the lengthening of pollen season.

How Does that Harm Our Health?

Poor air quality with an increase in ozone increases asthma attacks and can lead to disability, hospitalization, and death. Warmer and drier conditions lead to an increase in wildfires which fills the air with smoke (particulates). Fire smoke, which can travel hundreds of miles downwind, exposes people to harmful pollutants and increases emergency room visits and hospitalizations, and causes them to need treatments for asthma, bronchitis, pneumonia, chest pain, and other heart and lung conditions. **Figure 2-7** shows a child with asthma who is receiving treatment with an inhaler.

Figure 2-7 Poor air quality can cause worsening of asthma symptoms.
© xavier gallego morell/ShutterStock, Inc.

Warmer temperatures lead to a longer pollen season, and increased carbon dioxide in the air leads to higher pollen levels and stronger (more allergenic) pollen. These factors make allergies and asthma worse and more common. Higher humidity and flooding from heavy downpours can lead to dampness, and mold can grow indoors, which can also increase allergy symptoms. Since the majority of people with asthma have allergies, this can cause more severe asthma symptoms with all of the potential consequences.

Who Is Being Harmed?

Anyone can be negatively affected by poor air quality; however, people with preexisting chronic respiratory conditions such as asthma or chronic lung disease are extremely susceptible. Smoke from wildfires can push air quality into the danger zone. When this happens, risk is not limited to people who have preexisting respiratory conditions. Everyone is at risk, and some people are at even greater risk—especially children who breathe more times per minute and pregnant women who run the risk of premature labor.

The United States, especially the East Coast and Greater Los Angeles Area, has dangerously high levels of pollution, which lead to premature death, disability, and disease. In many urban areas, asthma rates are twice what they are nationwide. Longitudinal studies in southern California studying successive cohorts of children growing up as air quality was improving have demonstrated clearly that exposure to higher levels of pollutants leads to higher rates of asthma.[42]

Neurocognitive Effects of Air Pollution: The Very Young and the Elderly

Air pollution also contributes to **neurodevelopmental** damage to the growth and functioning of the brain and nervous system. Furthermore, new scientific research suggests that air pollution may also be a factor in **neurodegenerative** disorders that many older adults experience.[43]

Burning fossil fuels produces harmful air pollution and increases people's exposure to toxic chemicals that can harm their brains and nervous systems. The scientific community generally agrees that air pollution from fossil fuel use is harmful to children's developing brains and may also affect the cognitive functioning of older adults—although not all studies have found these results. In children, exposure to air pollution has been linked to developmental delays, reduced IQ, cognitive deficits, and autism spectrum disorder. In adults, exposure to air pollution has been linked to higher rates of dementia and Alzheimer's Disease. The very young, the elderly, and people with low household income are especially vulnerable to the harmful impacts of exposure to toxic chemicals in the air.

In 2016, *Environmental Health Perspectives* published a joint public statement issued by 14 scientific or medical associations and 50 scientists representing the disciplines of pediatrics, toxicology, public health, and neurobiology. The *Project TENDR* (Targeting Environmental Neurodevelopmental Risks) *Consensus Statement* noted evidence of danger to children in the United States due to air pollution, listing fossil fuel-related air pollutants (including particulate matter, polyaromatic hydrocarbons, and nitrogen dioxide) as "prime examples of toxic chemicals that can contribute to learning, behavioral, or intellectual impairment, as well as specific neurodevelopmental disorders such as ADHD [attention deficit hyperactivity disorder] or autism."[44]

In its 2017 report, the *Lancet Commission* on pollution and health stated that "pollution is now understood to be an important causative agent of many non-communicable diseases including … neurodevelopmental disorders." Specifically, it noted "emerging evidence" of causal associations from air pollution exposure to fine particulate matter and **decreased cognitive function, attention-deficit hyperactivity disorder** and **autism in children**, as well as **dementia in adults**.[45]

Fire, Lungs, and Hearts

By Dr. John Meredith, Emergency Room Physician, East Carolina University

In June 2008, a wildfire devastated Eastern North Carolina. The Evans Road Wildfire, which burned more than 45,000 acres and cost $20 million to battle, started in the midst of the state's worst drought. As the fire burned for three long months that summer, plumes of smoke carrying dangerous particles covered the eastern side of the state and beyond. **Figure 2-8** illustrates the tremendous smoke generated by such a fire.

Figure 2-8 Tremendous smoke can be generated by wildfires and can travel for hundreds of miles.

Figure 2-9 An Emergency Physician, Dr. John Meredith, saw patients in his Emergency Room who were affected by the fire.

Courtesy of John Meredith.

Researchers saw this fire as an opportunity to learn more about how harmful air pollution from wildfires is to human life. **Figure 2-9** is a picture of emergency room physician researcher, Dr. John Meredith. His team studied two sets of counties in North Carolina: those that were affected by the smoke and those that were not. See **Figure 2-10**. They tallied respiratory conditions including asthma, pneumonia, and common respiratory infections, as well heart attacks and other cardiac conditions.

Here are the results: Those living in counties affected by the plume had a 50% increase in the trips to emergency departments from respiratory illness like Chronic Obstructive Pulmonary Disease (COPD), pneumonia and bronchitis, while the other counties did not.[46] The smoke also caused a jump in emergency department visits for heart disease, including myocardial infarcts and heart failure, not seen in the other counties. People with heart disease are extremely sensitive to particles from wildfires.

Blazing forests are just as damaging to human health as they are to homes and neighborhoods. Increasing temperatures and more frequent droughts caused by climate change will increase the number of wildfires in the United States—and worldwide. Not only our lungs, but also our hearts, are at serious risk.

Air Pollution Around the World

Health problems stemming from air pollution are a global problem. The World Health Organization estimates that pollution of ambient air is responsible for 3.7 million deaths each year, with most of those deaths occurring in low- and middle-income countries. Most of the deaths are caused by heart, cerebrovascular, and lung diseases. A small percent is attributable to cancer.[47] In many lower income

Figure 2-10 Aerial map showing counties impacted by the Evans Road Fire at the Pocosin Lakes National Wildlife Refuge, North Carolina on June 10-12, 2008.

Rappold AG, et al. Peat bog wildfire smoke exposure in rural North Carolina is associated with cardiopulmonary emergency department visits assessed through syndromic surveillance. *Environ Health Perspect.* 2011;119(10):1415–1420. doi:10.1289/ehp.1003206.

countries, indoor use of wood or coal burning stoves exposes women (mostly) to unsafe levels of particulate pollution and contributes substantially to this mortality total. In 2016, more than 90% of the world's population was living in places where the air quality guidelines of the World Health Organization were not met.

Infectious Diseases
Ticks and Mosquitoes

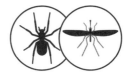

What Is Happening?

Climate change is causing increasing temperatures, too much or too little rain, and severe weather events that impact vector populations that carry disease.

How Does that Harm Our Health?

Along with the direct harms we've described above, these changes can lead to an increase in the number and geographic range of disease-carrying mosquitoes, fleas, and ticks.[48] Mosquitoes like that shown in **Figure 2-11** that carry diseases like West

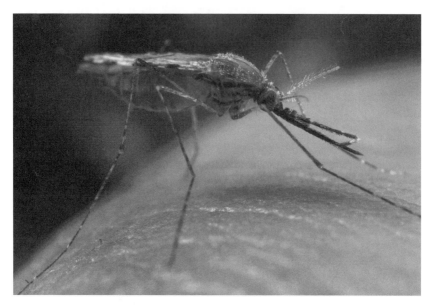

Figure 2-11 This is a photograph of a mosquito that is ready to pierce human skin.
© James Gathany/CDC.

Nile virus and dengue fever flourish in conditions that are becoming more common, and there is a concern that malaria may reemerge in the United States. Ticks that carry Lyme disease have become more numerous in many areas and have expanded their range northward and westward.[49] The tick that carries Lyme disease is reported in 45.7% of U.S. counties, up from 30% in 1998.[50] As insect carriers of infection move to new areas, diseases not typically found in those areas can spread.

Who Is Being Harmed?

Anyone can be harmed by these diseases, but people who spend more time outdoors—where these insects and other disease carriers live—are most vulnerable. This includes people who live in places with no window screens. Mosquitos also thrive in hotter, more humid locations or places where rain leaves standing water.

Four Seasons of Ticks and Mosquitoes

By Dr. Nitin Damle, President, American College of Physicians (ACP);
Founder, South County Internal Medicine, Inc.

© Dr. Nitin Damle

It is not a surprise that over the past five years, my practice has seen a rise in the incidence of tick-borne diseases, including Lyme disease and other infections. My physician colleagues used to treat two or three cases a month during tick season; now, each of us sees 40 to 50 new cases during each tick season.

Those blacklegged ticks, the carriers of Lyme disease, thrive in warm, muggy weather. In my home state of Rhode Island, where winters have gotten warmer and shorter, these tiny, sesame seed-sized insects have more time to bite humans and spread Lyme disease. **Figure 2-12** shows a fully engorged tick after biting a person or animal. Tick season used to be relegated to summer; it now spans spring and

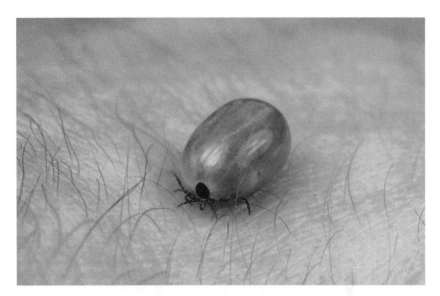

Figure 2-12 This is a photo of a tick that is fully engorged with blood after biting someone. Unengorged ticks look like more like apple pits or other seeds.

© Petr Jilek/ShutterStock, Inc.

autumn. And this isn't limited to the typical tick hotspot states. A picture of the characteristic rash of Lyme Disease may be found in **Figure 2-13**.

Across the country, doctors are seeing more patients struck ill by serious diseases like Lyme disease and West Nile fever. Because of the changing climate and the spread of vectors, we expect that U.S. residents will continue to face new diseases and familiar diseases in new places. I know that doctors need to be ready for this and patients need to understand these dangers. A picture of Dr. Damle may be found in **Figure 2-14**.

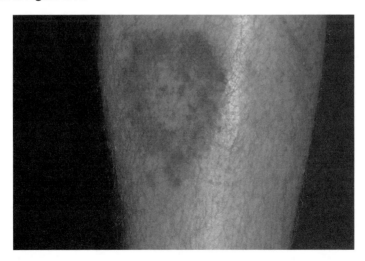

Figure 2-13 A bull's eye rash is characteristic of Lyme disease.

© E. M. Singletary, M.D. Used with permission.

Figure 2-14 Rhode Island internist, Dr. Nitin Damle, has seen an increase in people who have diseases caused by tick bites.

Dr. Nitin Damle.

Global Impacts of Insect Vectors—Dengue Fever, Malaria, Etc.

Disease occurrence is determined by a complex interplay of social and environmental conditions and can be affected by access to healthcare services. Some diseases are particularly sensitive to variations in climate and weather and may be affected by both longer-term climate changes and shorter-term extreme weather events. The extent of mosquito vectors, two species in particular—*Aedes aegypti* and *Aedes albopictus*—have increased since 1990. This has brought double the number of cases of dengue fever every decade since 1990.[51] This potential to transmit disease is referred to as "Vectorial capacity," and as explained in the Lancet Countdown report, it is growing.

Climate influences affect the spread of infectious diseases, and they often interact with other factors, including behavioral, demographic, socio-economic, topographic and other environmental factors. Understanding the contribution of climate change to infectious disease risk is thus complex, but an important part of understanding the impact of climate change on health. While global health initiatives have decreased deaths associated with climate-sensitive diseases since 1990, the current changes in the climate threaten to eradicate the progress made through public health initiatives over the last 50 years.[52]

Contaminated Water

What Is Happening?

Climate change is causing higher water temperatures, heavier downpours, rising sea levels, and, as a result, more flooding due to sea level rise and heavy downpours, each of which can lead to water contamination that makes people sick. In places where the run-off of storm water and sewage are combined in a common outflow track, a system overload can bring these types of water together. That mixture becomes the contaminated flood water that exposes people to sewage. Bodies of water can also become contaminated when heavy rains wash fertilizers off of farmland and into waterways.

How Does that Harm Our Health?

Each of these conditions can lead to contamination of drinking water, recreational water, fish, and shellfish—all of which can make people sick. For example, heavy rains can cause fertilizers and animal waste from farms to be flushed into rivers, lakes, and oceans. There, the excess nutrients and warm waters promote the growth of algae, viruses, parasites, and bacteria such as *Salmonella, Escherichia coli* and *Vibrio*. People are exposed to these pathogens by drinking or swimming in contaminated water or by eating contaminated fish and shellfish. This can cause diarrhea and vomiting and, in severe cases, paralysis, organ failure, and death.[53] A 1993 *Cryptosporidium* outbreak in Milwaukee, which made more than 400,000 people ill, coincided with record high flows in the Milwaukee River, a reflection of the amount of rainfall in the watershed.[54] The toxic algae bloom in Lake Erie that required shutting down the water supply in Toledo, Ohio, in 2014, is another example. See **Figure 2-15**.

Figure 2-15 A couple paddle surfs as algae surfaces on Lake Erie in August 2014. Health officials sent samples to several laboratories for testing after finding Lake Erie was affected by a "harmful algal bloom." The lake provides the bulk of the area's drinking water.

Who Is Being Harmed?

Anyone can be harmed by contaminated water, but some people—especially children, the elderly, people with weakened immune systems, people in remote or low-income communities with inadequate water systems, and people in communities who live close to the water's edge or whose livelihoods depend on fish and shellfish—are at higher risk.

Global Burden of Water Contamination

Water is another main way that populations feel the effects of climate change. Water availability is becoming less predictable in many places, and increased incidences of flooding threaten to destroy water points and sanitation facilities and contaminate water sources.

Access to clean water is already a substantial global problem; 2 billion people don't have basic sanitation facilities. Climate change not only threatens access to clean water but also access to any water in many areas. Currently, 2 billion people live in countries that experience high water stress. This is expected to get worse as populations grow and their need for water increases. As climate change intensifies, this is likely to deteriorate further over time.[55] Higher temperatures and more extreme, less predictable weather conditions will affect availability and distribution of rainfall, snowmelt, groundwater and rivers, and further reduce water quality. Low-income communities, which are already the most vulnerable to threats to the water supply, will be most affected. In addition, loss of glaciers in the Andes Mountains and the Himalayas threatens the water supply for billions of people who live on the land masses of South America and the Indian subcontinent.[56] In some regions, droughts are making existing water scarcity worse and negatively impacting people's health and productivity. Ensuring that everyone has access to sustainable water and sanitation services is a critical protective climate change adaptation strategy for years to come.[57]

Contaminated Food, Food Scarcity, Declining Nutrition

What Is Happening?

Climate change is causing increases in temperature, humidity, and extreme weather events like droughts, heavy downpours, and flooding. This can cause contamination of food or loss of crops, which threatens the food supply.

How Does that Harm Our Health?

Food Contamination. Each of these conditions can lead to food becoming contaminated by bacteria and toxins. For example, heavy downpours and flooding can spread fecal bacteria and viruses into fields where food is growing. See **Figures 2-16** and **2-17**. Higher sea surface temperatures can lead to more bacterial pathogens. In

Figure 2-16 A flooded cornfield in Gladstone, Illinois.

© J. J. Gouin/Shutterstock.

some cases, drought can lead to a greater concentration of bacteria in waterways. Foodborne illness has long been known to peak in summer when the temperatures are higher. Because pests, parasites, and bacteria thrive in warmer temperatures, farmers are using more pesticides on crops and drugs in livestock, which can cause health problems. The geographic range of mold and associated toxins is also expanding, affecting corn, peanuts, cereal grains, and fruit.[58]

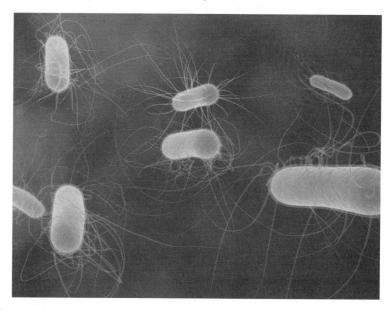

Figure 2-17 Coliform bacteria can contaminate waterways and make people sick if ingested.

© Sebastian Kaulitzki/ShutterStock, Inc.

Nutritional Value. Rising levels of carbon dioxide in the air decrease the nutritional value of important food crops such as wheat, rice, barley, and potatoes.[59] This happens because in the presence of more carbon dioxide, these plants produce less protein and more starch and sugar, and they take in fewer essential minerals. Rice is more likely to undergo these changes than other grains.[60] Higher temperatures can also result in more food spoiling. Drought can damage or destroy crops, and extreme weather events can disrupt food production and distribution by knocking out power, damaging infrastructure, and delaying food shipments. As a result, food can be damaged, spoiled, or contaminated, reducing the availability of and access to safe and nutritious food.[61]

Who Is Being Harmed?

Anyone can be harmed by contaminated food, but infants, young children, pregnant women, the elderly, the poor, agricultural workers, and those with weakened immune systems are more susceptible. Decreased food supply or food scarcity raises the cost of food, making it more difficult for those who are food insecure.

Worldwide Threats to Nutrition and Food Insecurity

A United Nations report released in July 2019 made clear that the world is not on track to end extreme poverty by 2030, the target date for the current United Nations Sustainable Development Goals, and that hunger is growing.[62] A recent UN report made clear that progress is being undermined by the impact of climate change along with increasing inequality.[63] Increasing temperatures on the planet and more variable rainfalls are expected to reduce crop yields in many tropical developing regions where food security is already a problem.[64] More floods and severe droughts are predicted. Changes in water availability will impact health and food security and have already been proven to trigger refugee dynamics and political instability. The rise in ocean temperatures and the acidification that accompanies this, along with other human and environmental factors, is depleting the fish and seafood supply. Fish and seafood are staples of the diet for many peoples.

Conflict over resources will occur among nations. Conflict over fishing rights and boundaries is frequently in the news. Food insecurity can cause climate migration and conflict. The migration itself becomes a cause of conflict. Loss of crops related to climate change is an explanation for the increased migration from Central America that is so politically contentious in the United States.[65] Initial protests against the Syrian government were launched from a region that faced food shortages due to drought.[66] Serious conflicts and civil wars have already started to grow out of these difficult human challenges.

Threats to Mental Health

© Elena Kalinicheva/Shutterstock.

What Is Happening?

Climate change is causing damage and displacement to people's homes and lives due to increases in the extreme weather events such as heavy downpours, floods, droughts, and major storms. In addition to the physical harms described above, these events can cause psychological trauma that impacts mental health.

How Does that Harm Our Mental Health?

Many people exposed to the worst extreme weather events, especially those that disrupt their homes, incomes, and social networks, experience stress and serious mental health issues including depression, anxiety, post-traumatic stress disorder (PTSD), and increases in suicidal thoughts and behavior. Such disasters are also associated with increases in alcohol use, drug abuse, or domestic violence. See **Figures 2-18** and **2-19** which offer pictures of distressed people receiving assistance after Hurricane Sandy in New York in 2012. Children may experience prolonged separation from their parents. Beyond the well-known risks that specific disasters pose to our mental health, the physical, social, and economic stresses created by climate change all increase our risk of mental health problems.[61]

Who Is Being Harmed?

Anyone's mental health can be harmed by a disaster, but people at higher risk include children, the elderly, pregnant women, those with preexisting mental illness, the poor, homeless, and first responders. It took six months after Hurricane Katrina for all of the children who were separated from their parents to be reunited with them.[67] Such separation qualified as an adverse childhood event. Children who experience adverse childhood events have been shown to have poorer health throughout their lives.[68] Farmers and other people who rely on the natural environment for their livelihoods are also at higher risk.

Global Implications: National and International Conflicts

As a feature of the all-encompassing environment in which humans dwell, it does not seem surprising that environmental temperature should have an impact on human behavior. The global implications of climate change for mental health are powerful. Heat is known to potentiate violence as documented in

Figure 2-18 People on a food line after Hurricane Sandy.
U.S. Department of Defense.

Figure 2-19 Neighbors embrace after looking through the wreckage of their homes devastated by fire and the effects of Hurricane Sandy in Queens, NY on October 31, 2012.

© John Moore/Getty Images News/Getty Images.

the psychiatric literature.[69,70] The "long hot summers" of increased violence and rioting faced by urban centers in the United States were the stimulus for the free festivals and entertainment in the public parks that have become a staple of urban summers. Recent research had identified a rise in the suicide rate associated with higher temperature.[71]

Individual mental health impacts can result from climate change-related health issues and personal losses such as exacerbation of chronic diseases or new chronic diseases, deaths, displacement, loss of homes and property, etc. There are also larger population-based dangers resulting from food insecurity because of droughts and famines. Although predicted for years by climate health experts, migration is an obvious result. And where people are on the move in large numbers due to droughts and food insecurity, there is a significant risk of conflict over resources that may lead to exploitation or violence. Recent examples illustrate this.

A drought in a region of Syria led to demands by the affected population for assistance from the government of Bashir Assad. The government accused the people of disloyalty and insurrection; serious conflict developed from there. The war that resulted caused millions of Syrian nationals to flee to Lebanon, Turkey, and Europe. The presence of a group of refugees of this size caused dissension in Europe and political conflicts developed, which focused on migration and entry of foreign nationals.

The struggle over immigration that developed in the United States in recent years has focused on people seeking refuge from Mexico and Central America. Recent news has been about "caravans'" of people from Central America. A drought and losses by farmers unable to make a living or feed their families are factors at the root of this surge of immigration. Such conflicts are bringing the developed world, where the capacity to compensate for disasters is greater, into conflict with the developing world of low- and middle-income countries.

Table 2-1 Percent of Physicians Who Respond Their Patients Are Affected by Climate Change

Physicians in three medical societies* were asked the following question: In which of the following ways, if any, do you think your patients were being affected by climate change in 2014 and 2015 (N = 1868 - 1908)?[3,5,10]

Health Harms	Responded "Yes"
Air pollution-related	76%
Allergic symptoms	63%
Injuries due to storms, etc.	57%
Heat effects	45%
Vector-borne infections (e.g., infections spread by mosquitoes or ticks)	40%
Diarrhea from food or water infections	29%
Mental health**	40%

*The societies are the American Academy of Allergy, Asthma, Immunology; the American Thoracic Society; and the National Medical Association.
**Results from one medical society only.
Sarfaty M, Gould R, Maibach E. *Medical alert! Climate change is harming our health.* Fairfax, VA: Medical Society Consortium on Climate and Health; 2017.

What the Public and Doctors Think About Climate Change and Health

> **Fact:** Based on several surveys, two of every three doctors think climate change has direct relevance right now to patient care.[3,5,10]
>
> **Fact:** Physicians say the most common ways that climate change is harming their patients' health are through poor air quality, worsening allergies, injuries due to storms, heat-related illness, and infections spread by mosquitoes and ticks.[3,5,10] See **Table 2-1**.

© The Medical Society Consortium on Climate and Health.

Risk: The Climate Future Health Professionals Worry About

In the report *What We Know*, provided by the American Association for the Advancement of Science (the largest science organization in the United States), climate scientists concluded:

> *We are at risk of pushing our climate system toward abrupt, unpredictable, and potentially irreversible changes with highly damaging impacts.*[5]

Figure 2-20 A flooded street along the water on the island of Manhattan, NYC, after a storm.

© STAN HONDA/Staff/Getty Images.

The scientists tell us that, even if we somehow stopped adding greenhouse gases to our atmosphere tomorrow, more warming is now "baked in" to our climate. But the greatest risk is that at some point, warming will cause abrupt and irreversible changes. Such scenarios include the large-scale collapse of ice sheets in Antarctica, creating the potential for 23 feet of rise in sea level.[6] It is disturbing that scientists cannot tell us how much warming it would take to trigger such scenarios. **Figure 2-20** shows the encroachment of water in New York City along a waterfront street as the temperature continues to rise.

As one health professional put it, "Even our 'best-case scenario' means we're going to be seeing more [people] with demanding health problems. But the worst-case scenarios of climate change really worry me. It would mean a level of human suffering we can barely contemplate, much less respond to."[72,73]

What We Can Do: Prepare and Prevent

One of the key findings in the 2014 and 2018 National Climate Assessments—the most comprehensive assessments on the impacts of climate change in the United States—was the following:

> *Public health actions, especially preparedness and prevention, can do much to protect people from some of the impacts of climate change. Early action provides the largest health benefits. As threats increase, our ability to adapt to future changes may be limited.*

Prevention is a central tenet of public health. Many conditions that are difficult and costly to treat when a patient gets to the doctor could be prevented before they occur at a fraction of the cost. Similarly, many of the larger health impacts associated with climate change can be prevented through early action at significantly lower cost than dealing with them after they occur.[74]

—The U.S. National Climate Assessment 2014

Doctors agree with climate scientists: the sooner we take action, the more harm we can prevent, and the more we can protect the health of all U.S. residents.[73]

WRAP-UP

Discussion Questions

1. Is everyone equally likely to be negatively affected by climate change? Who is likely to experience greater harm and why?
2. Discuss several ways that climate change endangers the health of children specifically.
3. Discuss several ways that increased flooding can negatively impact human health.
4. Discuss how public health actions, especially preparedness and prevention, can protect people from some of the impacts of climate change, especially if action is taken early.

References

1. Maibach EW, Kreslake JM, Roser-Renouf C, Rosenthal S, Feinberg G, Leiserowitz AA. Do Americans understand that global warming is harmful to human health? Evidence from a national survey. *Ann Global Health*. 2015;81(3):396-409.
2. Leiserowitz A, Maibach E, Roser-Renouf C., et al. (2017). Climate change in the American mind: November 2016. Yale University and George Mason University. New Haven, CT: Yale Program on Climate Change Communication.
3. Sarfaty M, Kreslake J, Casale T, Maibach EW. Views of AAAAI members on climate change and health. *J Allergy Clin Immunol-In-Practice*. 2015;4(2):P333-P335.
4. Koh H. Communicating the health effects of climate change. *JAMA*. 2016;315(3):239-240.
5. Sarfaty M, Bloodhart B, Ewart G, et al. American Thoracic Society member survey on climate change and health. *Ann American Thoracic Soc*. 2015;12(2):274-278.
6. Balbus J, Crimmins Gamble JL, et al.: The Impacts of Climate Cahnge on Human Health in the United States: A Scientific Assessment. Crimmins, Global Change Research Program, Washington, DC, 2016;312.
7. Wellbery C, Sarfaty M. The health hazards of air pollution—implications for your patients. *Am Fam Physician*. 2017;95(3):146-148.
8. Crowley RA. Climate change and health: a position paper of the American College of Physicians. Health and Public Policy Committee of the American College of Physicians. *Ann Intern Med*. 2016;164(9):608-610.
9. Ahdoot S, Pacheco SE. Council on Environmental Health. Global Climate Change and Children's Health. *Pediatrics*. 2015;136(5):e1468-e1484.

10. Sarfaty M, Mitchell M, Bloodhart B, Maibach E. A survey of African American physicians on the health effects of climate change. *Int J Environ Res Public Health*. 2014;11(12):12473-12485.

11. Policy of the American Medical Association, 2008 reaffirmed 2014; H-135.938 Global Climate Change and Human Health; https://searchpf.ama-assn.org/SearchML/policyFinderPages /search.action

12. Policy of the American Medical Association, 2016; H-135.923; AMA Advocacy for Environmental Sustainability and Climate; https:// searchpf.ama- assn.org/SearchML/ searchDetails.action? uri=%2 FAMADoc%2 FHOD- 135.923.xml

13. Climate Change in the American Mind, April 2019 and April 2020. https://www.climate changecommunication.org/wp-content/uploads/2019/06/Climate_Change_American _Mind_April_2019b.pdf. Accessed September 3, 2019 and https://www.climatechange communication.org/wp-content/uploads/2020/05/climate-change-american-mind-april -2020b.pdf: page 18. Accessed August 7, 2020.

14. Gamble JL, Balbus J, Berger K, et al. Ch. 9: Populations of Concern. The Impacts of Climate Change on Human Health in the United States: A Scientific Assessment. U.S. Global Change Research Program, Washington, DC, 247-286.

15. https://climatecommunication.yale.edu/publications/public-perceptions-of-the-health -consequences-of-global-warming/. Accessed February 19, 2017.

16. Leiserowitz A, Maibach E, Roser-Renouf C, et al. Public perceptions of the health consequences of global warming: October, 2014. New Haven: Yale Project on Climate Change Communication. 2014.

17. Leiserowitz, A., Maibach EW. Climate Change in the American Mind. November, 2011. Yale Program on Climate Communication and George Mason University Center for Climate Change Communication.

18. Gallup Poll 2014.

19. Krygsman K, Speiser M, Lake C, Voss, J. (2017). American Climate Metrics Survey 2016: National. ecoAmerica and Lake Research Partners. Washington, D.C.

20. Brenan M. Nurses again outpace other professions for honesty, ethics. *Gallup*. 2018. https:// news.gallup.com/poll/245597/nurses-again-outpace-professions-honesty-ethics.aspx. Accessed September 3, 2019.

21. Wang M, Aaron CP, Madrigano J, et al. Association between long-term exposure to ambient air pollution and change in quantitatively assessed emphysema and lung function ... — JAMA, 2019 — jamanetwork.com

22. Bekkar B, Pacheco S, Basu R, DeNicola N. Association of Air Pollution and Heat Exposure With Preterm Birth, Low Birth Weight, and Stillbirth in the US: A Systematic Review. *JAMA Netw Open*. 2020;3(6):e208243. doi:10.1001/jamanetworkopen.2020.8243

23. Crimmins A, Balbus J, Gamble JL, et al. USGCRP, 2016: The Impacts of climate change on human health in the United States: A scientific assessment. Global *Change Research Program*, Washington, DC, 312 pp. http://dx.doi. org/10.7930/J0R49NQX.

24. Centers for Disease Control and Prevention (CDC). Nonfatal sports and recreation heat illness treated in hospital emergency departments: United States, 2001–2009. *MMWR Morb Mortal Wkly Rep*. 2011;60(29):977-980.

25. Centers for Disease Control and Prevention (CDC). Heat illness among high school athletes: United States, 2005-2009. *MMWR Morb Mortal Wkly Rep*. 2010;59(32):1009-1013.

26. Ibid.

27. Lancet Count Down 2017.

28. Sorenson C, Garcia-Trabanino R. A New era of climate medicine—addressing heat-triggered renal disease. *N Engl J Med*. 2019;381:693-696. https://www.nejm.org/doi/full/10.1056 /NEJMp1907859

29. Glaser J, Lemery J, Rajagopalan B, et al. Climate change and the emergent epidemic of CKD from heat stress in rural communities: the case for heat stress nephropathy. *Clin J Am Soc Nephrol* 2016;11(8):1472-1483.

30. https://edition.cnn.com/2015/06/01/asia/india-heat-wave-deaths/index.html

31. Read in TIME: https://apple.news/ARTz97WerQRWt1ldH0C2jTw. Accessed September 14, 2019.

32. https://www.sun-sentinel.com/local/broward/fl-reg-nursing-home-dehydration-20171227-story. html. Accessed December 23, 2012.

33. "Hurricane Dorian Becomes the 5th Atlantic Category 5 in 4 Years." *The Weather Channel*. Retrieved September 13, 2019.

34. https://psmag.com/news/americas-latest-500-year-rainstorm-is-underway-right-now-in-louisiana. Accessed December 23, 2019.

35. https://en.wikipedia.org/wiki/2016_ Louisiana_floods. Accessed February 20, 2017.

36. Watts N, Amann M, Arnell N, et al. The 2019 report of *The Lancet* Count Down on health and climate change: ensuring that the health of a child born today is not defined by a changing climate. Lancet. 2019;394(10211):P1836-P1878.

37. The Lancet. Lancet Count Down 2017. https://www.thelancet.com/doi/story/10.1016/vid.2017.10.27.6156

38. Ferris R. US disaster costs shatter records in 2017, the thrid-warmest year on record. *CNBC*. 2018. https://www.cnbc.com/2018/01/08/us-disaster-costs-shatter-records-in-2017-the-third-warmest-year.html. Accessed December 23, 2019.

39. NASA Earth Observatory. A devastating stall by Hurricane Dorian. 2019. https://earthobservatory.nasa.gov/images/145559/a-devastating-stall-by-hurricane-dorian. Accessed September 14, 2019.

40. Watts NA, et al. The 2019 report of The Lancet Countdown on health and climate change: ensuring that the health of a child born today is not defined by a changing climate November 13, 2019. 394(10211):Section 1.2.3.

41. Hu P, Zhang Q, Shi P, Chen B, Fang J. Flood-induced mortality across the globe: spatiotemporal pattern and influencing factors. *Sci Total Envir*. 2018;643:171-182.

42. Gauderman WJ, Avol E, Gilliland F, et al. The effect of air pollution on lung development from 10 to 18 years of age. *N Engl J Med*. 2004:351:1057-1067.

43. Heusinkveld HJ, Wahle T, Campbell A., et al. Neurodegenerative and Neurological Disorders by Small Inhaled Particles. *Neurotoxicology*. 2016 Sep;56:94-106. Epub 2016 Jul 19.

44. Bennett D, Bellinger DC, Birnbaum LS, et al. Project TENDR: Targeting Environmental Neuro-Developmental Risks The TENDR Consensus Statement. *Environ Health Perspect*. 2016;124(7):A118-A122. doi:10.1289/EHP358

45. The Lancet. The Lancet Commission on pollution and health. *Lancet*. 2017. http://www.thelancet.com/commissions/pollution-and-health.

46. Sarfaty M, Gould R, Maibach E. *Medical alert! Climate change is harming our health*. Fairfax, VA: Medical Society Consortium on Climate and Health; 2017.

47. World Health Organization (WHO). Ambient (outdoor) air pollution. https://www.who.int/news-room/fact-sheets/detail/ambient-(outdoor)-air-quality-and-health. Accessed May 22, 2020.

48. Beard CB, Eisen RJ, Barker CM, et.al. Vectorborne Diseases. The Impacts of Climate Change on Human Health in the United States 2016: A Scientific Assessment. U.S. Global Change Research Program, Washington, DC, 129-156. http://dx.doi.org/10.7930/J0765C7V https://s3.amazonaws.com/climatehealth2016/low/ClimateHealth2016_05_Vector_small.pdf. Accessed February 24, 2017.

49. https://www.lymediseaseassociation.org/resources/cases-a-other-statistics. Accessed February 24, 2017.

50. Eisen R, Eisen L, Beard CB. County-scale distribution of Ixodes scapularis and Ixodes pacificus (Acari: Ixodidae) in the continental United States. *J Med Entomol*. 2016;53(2):349-386.

51. Watts N, Amann M, Ayeb-Karlsson S, et al. The *Lancet* Countdown on health and climate change: from 25 years of inaction to a global transformation for public health. *Lancet*. 2017;391(10120):P581-P630.

52. Ibid.

53. Trtanj J, Jantarasami L, et. al. Climate Impacts on Water Related Illness Chapter in Impacts of Climate Change in the United States: A Scientific Assessment. 2016. Global Change Research Program. https://s3.amazonaws.com/climatehealth2016/low/ClimateHealth2016_06_Water_small.pdf. Accessed February 24, 2017.

54. Patz JA, Vavrus SJ, Uejio CK, McLellan SL. Climate change and waterborne disease risk in the Great Lakes region of the US. *A J Prevent Med*. 2008;35:451-458.

55. United Nations. Water and climate change. (n.d). https://www.unwater.org/water-facts/climate-change/. Accessed September 14, 2019.

56. McCarthy J, Sanchez E. Billions rely on Himalayan glaciers for water. But they're disappearing. *Global Citizen*. 2019. https://www.globalcitizen.org/en/content/himalayas-melting-climate-change/. Accessed September 14, 2019.

57. United Nations. Water and climate change. 2017. https://www.unwater.org/water-facts/climate -change/

58. Ziska L, Crimmins A, Auclair A, et al. Ch. 7: Food Safety, Nutrition, and Distribution. The impacts of climate change on human health in the United States: A scientific assessment. *U.S. Global Change Research Program*, Washington, DC, 189-216.

59. Sustainable Development Goals. Knowledge Platform. Transforming our world: the 2030 agenda for sustainable development. (n.d). https://sustainabledevelopment.un.org/post2015 /transformingourworld. Accessed September 14, 2019.

60. Ibid.

61. Dodgen D, Donato D, Kelly N, et al. Mental health and well-being. The impacts of climate change on human health in the United States 2016: a scientific assessment. *U.S. Global Change Research Program*, Washington, DC, 217-246.

62. Lederer EM. UN: Climate change undercutting work to end poverty, hunger. Associated Press. July 10, 2019.

63. https://www.apnews.com/88aa7e13071e427d9ef15ea2ac996f82. Accessed September 14, 2019.

64. Zhu C, et al. Carbon dioxide (CO_2) levels this century will alter the protein, micronutrients, and vitamin content of rice grains with potential health consequences for the poorest rice-dependent countries. Science Advances 23 May 2018;4(5):eaaq1012.

65. Hallett MC. How climate change is driving emigration from Central America. 2019. https:// www.pbs.org/newshour/world/how-climate-change-is-driving-emigration-from-central -america. September 18, 2019. Accessed February 20, 2020.

66. Kelley CP, Mohtadi S, Cane MA, Seager R, Kushnir Y. Climate change in the Fertile Crescent and implications of the recent Syrian drought. *Proc Natl Acad Sci U S A.* 2015;112(11): 3241-3246.

67. Reckdahl K. The Lost Children of Katrina. The Atlantic April 2, 2015. https://www.theatlantic. com/education/archive/2015/04/the-lost-children-of-katrina/389345/. Accessed February 20, 2020.

68. Jack P, Shonkoff JP, Garner AS. Report of the committee on psychosocial aspects of child and family health, committee on early childhood, adoption, and dependent care, and section on developmental and behavioral pediatrics, Siegel BS, Dobbins MI, Earls MF, et al. *Pediatrics* 2012;129(1):e232-e246.

69. Anderson CA. Heat and violence. *Current Directions in Psychological Science.* 2001;10(1):33-38.

70. Berry HL, Bowen K, Kjellstrom T. Climate change and mental health: a causal pathways framework. *Int J Public Health.* 2010;55(2):123-132.

71. Burke M, González F, Baylis P, et al. Higher temperatures increase suicide rates in the United States and Mexico. *Nature Climate Change* 2018;8:723-729.

72. http://nca2014.globalchange.gov/report/sectors/human-health#statement-16522

73. American Association for the Advancement of Science. What We Know: The Risks, Reality, and Responses to Climate Change. Page 8. http://whatweknow.aaas.org/wp-content/uploads /2014/07/whatweknow_website.pdf. Accessed February 19, 2017.

74. 2014 National Climate Assessment. Global Change Research Program. http://nca2014. globalchange.gov/report

CHAPTER 3

Communicating About Health and Climate Change

Simple clear messages, repeated often, by a variety of trusted voices.

Edward W. Maibach, PhD,
George Mason Distinguished University Professor

KEY TERMS

Anthropogenic
Co-benefits
Draw Down
Health Frame

Segmentation
Self-efficacy
Sequester

CHAPTER OBJECTIVES

1. Explain why health is an effective way to speak to people about climate change.
2. Describe the characteristic of each of the groups of the "The Six Americas."
3. List several frames for communicating about climate change.
4. Describe the health benefits of solutions to climate change.
5. Summarize the advice of the communication experts about how to communicate effectively and affect people's thinking.

How to Talk About Climate Change and Health

Climate change and the many ways that it affects human health is made complex by widespread misinformation, so speaking about climate change may seem like an overwhelming task. Misinformation has been perpetrated by institutions such as the Heartland Institute, for which scientific accuracy is not a priority. Their website states quite clearly that, "We believe ideas matter, and the most important idea in human history is freedom."[1] Much of the funding for this group comes from sources that are involved with fossil fuel businesses or investments. That is where their profits originate.[2] In the 1990s, Heartland worked with the tobacco company Philip Morris to raise doubts about whether secondhand smoke posed cancer risks, and to lobby against government public health regulations. In 2008, the Institute started organizing public meetings to question the scientific consensus on climate change.

Some people claim that global warming doesn't exist or that it is happening but is not caused by human actions; others claim that it isn't so bad, and that there is nothing that can be done about it even if it is happening. As such, information without a scientific basis that is motivated by economic "freedom" should not stand unchallenged in this scientific era. A direct approach to presenting the facts about what is happening and personal stories of how it affects health can help to eliminate such obstacles to recognizing reality. Uncovering the basis for their beliefs ("What makes you think that this is the case?") and probing areas of doubt ("What would you like to understand better?") while encouraging critical thinking is also valuable.

In this chapter, we will discuss many useful approaches to make your communication more effective. We offer tips and simple strategies for successful communications regarding climate change and health, as well as key messages about each health risk that offer a foundation for communicating. This chapter also includes frequently asked questions. The truth is that human-caused climate change is occurring now, it is already harming human health, it will continue to worsen if we don't do something about it, and we can still lessen the impact if we act now. It is not too late yet. As health professionals, armed with effective communication tools and knowledge, you have a responsibility to share this information with the communities you serve, helping those who are at the greatest risk to protect themselves from the harmful effects and to prevent the atmosphere from degrading further with all of the negative health consequences that degradation will bring.

Climate Change from a Public Health Perspective

Current and anticipated health threats from climate change require public health action. Public health consists of several essential functions that can be seen in the accompanying figure.[3] **Figure 3-1** shows the public health wheel that summarizes public health professionals' essential functions on all issues that impact health. These functions include several categories of activity, namely assessment, policy development, and assurance. Assessment must include monitoring the prevalence and spread of disease, investigating outbreaks, and researching health effects. Policy development includes developing preparedness plans. Assurance means getting people the public health services they need through workforce development and providing or linking to medical care, and monitoring for whether needs are met.

Predictive modeling is basic to these efforts. Predictive modeling enables us to formulate an idea of the conditions a community will face and calculate how preparedness planning can make a difference. Another critical role for public health practitioners is informing policy makers about current and future risks and how they can help a community to alleviate and adapt to climate change, including the resources required.

Informing the public and providing essential information for self-protection as well as resources for community protection are also critical roles. This chapter addresses the communication role of "Inform, Educate, Empower." Communication connects public health professionals and the world as they provide the essential public health services.[4,5]

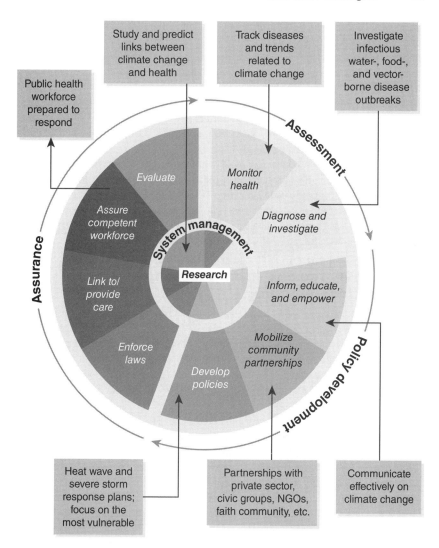

Figure 3-1 Public Health Actions on Climate Change Are Offered in the Boxes Outside the Circle.

Michael McGeehin, PhD. National Center for Environmental Health, CDC.

Communication Strategies

Much public health work is connected to health status or health conditions such as obesity or diabetes. Some begins with a behavioral issue such as smoking or not getting vaccinated. Global warming is neither a human diagnosis nor a behavior—it is based on a change in our atmosphere and on our reliance on burning fuel for energy—but it has direct health consequences. Regardless of the type of health danger, the work of educating people about health risks from global warming (also known as climate change) and empowering them to protect themselves and prevent or alleviate the impact is similar to other public health challenges.

Experience with disease prevention and health promotion challenges such as human immunodeficiency virus (HIV), smoking, and physical activity has helped establish effective health communication strategies. There are many communication models and theories to consider, including the social cognitive model, health belief model, precaution adoption principle model, stages of change, social marketing, and others.[6] The social cognitive model emphasizes the importance of what people believe about the thoughts or behavior of others in their social group or other social groups. The health belief model stresses the importance of influential factors on an individual's thinking. Precaution adoption stresses deliberate decision making based on newly acquired knowledge. Social marketing posits that ideas are acceptable if they are marketed (promoted).

Communication strategies may or may not be based on behavior change models, while some campaigns are based on pieces of several different theories.[7] Successful campaigns rely on communication research to demonstrate the efficacy of their strategies. An evidence base is essential to messages that move people toward solutions that protect and enhance their own health and the health of their communities. Research is helping communication about global warming to progress.

Media Coverage and Its Impact to Climate Change Understanding

The majority of Americans are *aware* that climate change is a problem, and increasing numbers are reportedly worried about it "a great deal."[8] However, despite the growing concern, the public doesn't understand enough about climate change. Less than half of the population understands that climate change is harming people in the United States now.[9]

Additionally, public confidence in the media's credibility when it comes to climate change has varied widely, which compounds this issue. Communication failures between scientists and journalists often lead to a significant discrepancy between the scientific and media presentation of **anthropogenic** climate change.[10] For example, according to a content analysis of reputable US newspapers, while some people believe that the media presents accurate information, nearly equal proportions believe that it is either exaggerated or underestimated.[11] Investigating the gaps between scientific knowledge, media reports, and public understanding illustrates the need for accurate information to be shared by credible and trusted sources, such as health professionals.

A specific decline in accurate media coverage occurred due to specific factors. The first was related to a phony scandal in which there were accusations, subsequently proven false, of scientific misconduct in a British research program. Another undermining influence occurred as the result of industry-funded efforts to plant the seeds of doubt about the accuracy of climate change science. For several years, several media outlets believed false equivalency and presented "both sides" of the climate issue. They believed (incorrectly) that there was a lot of scientific debate about the reality of climate change. This impression was publicized by several people with negative intentions.

There has not been any serious scientific debate about the existence of climate change presented in the peer reviewed literature. More than 97% of the publications, and of the climate scientists who have published, have concluded that climate change is real and that humans are the source of the problem. No credible analysis exists that suggests an alternative to the conclusion that climate change is caused by humans.

Americans' attitudes regarding climate change have been studied over time by two major climate communication centers. The Center for Climate Change Communication at George Mason University in partnership with the Yale Project on Climate Change Communication began biannual representative national surveys of American adults in 2008 as a part of the *Climate Change in the American Mind* project. This work examined the thinking of tens of thousands of people. Because of this work, Americans' understanding and beliefs about climate change have been summarized, and it is clearly understood that they want to address the global warming threat. Six groups have been characterized. They are often referred to as "Global Warming's Six Americas."

Global Warming's Six Americas

When examining large numbers of people in representative surveys over a decade, the population can be divided into six different groups based on perception of climate change. People in each of the six groups—*Global Warming's Six Americas*—have a distinct set of beliefs, behaviors, policy preferences, and engagement with the issues surrounding global warming.[12] The six groups are described below with the percentage of the U.S. population represented by each as of 2018 (**Figure 3-2**).

The Alarmed (29%): Members of this segment are heavily involved with global warming. They understand that it is happening, that it is caused by human activity, and that it is a very real threat. They are changing their behaviors and strongly support vigorous national policies to address global warming.[13]

The Concerned (30%): Members of this segment are similar to members of The Alarmed, except they are less certain of their conclusions, less personally engaged, less likely to change their behaviors, and slightly less supportive of policies to address climate change.

The Cautious (17%): Members of this segment tend to believe that global warming is happening, but they are less certain that humans caused global warming and less certain that the consequences will be serious. They do not consider global warming to be a personal threat. Also, they will probably not be willing to make personal behavior changes. However, they do show moderate support for policies that address climate change.

The Disengaged (5%): Members of this segment do not give much thought to global warming. They generally respond "Don't know" to most survey questions

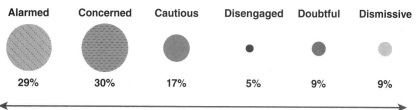

Figure 3-2 Global Warming's "Six Americas."

Reproduced from Leiserowitz A, Maibach E, Roser-Renouf C, Feinberg G, Rosenthal S. *Global Warming's Six Americas, March 2015.* Yale University and George Mason University. New Haven, CT: Yale Program on Climate Change Communication; 2015.

about global warming. They are inclined to believe that if global warming is real, it is likely harmful. They show moderate support for climate change policies and they acknowledge that they could easily change their minds about global warming.

The Doubtful (9%): About one-third of the members of this segment think that global warming is happening but that it is not caused by people. Another one-third say they do not know if it is happening. And the remaining one-third think it is not happening. None of these people considers global warming a threat now or in the foreseeable future. And almost all of them believe that the United States is already doing enough to address global warming.

The Dismissive (9%): Relatively speaking, members of this segment are highly engaged in global warming as an issue. They consider themselves very well informed about it, and they believe that it doesn't exist, and therefore it is not a threat. As a result, they feel strongly that policies to address global warming are misguided.

While there are minor differences among the members of these segments in demographic characteristics, such as age and gender, the most profound differences are in political ideology and party affiliation. Up to this point, the difference in political philosophy has been challenging for public involvement and public policy.[14] Regardless, **segmentation** *analysis* has made it possible to develop and test different communication strategies and messages for different groups.[15] This research has enabled use of evidence-based strategies when communicating about global warming.

The benefit of segmenting a population, as in the Climate Change in the American Mind project, is that targeted messaging can be developed for specific segments of the population. For those who are already "Alarmed," and eager to take action, offering suggestions about what they can do will be helpful. This group is ready to take action. Getting them involved with a productive activity that has an impact on problem solving in their community, state, and country is a very positive thing. They may become voices for a better future and help make a compelling case to others that, for the sake of everyone's health, implementing solutions is vital. If they have not already received training, they will probably benefit from learning how to take action, including letter writing, visiting policymakers, presenting their views to the media and social media, and how to work with others to create a consistent and unified grassroots solution. They should also learn what not to do.

For those who are "Concerned," a similar approach will work. This group may need to be encouraged to get additional training (similar to what was prescribed for The "Alarmed.") They should be persuaded to attend events and become involved in planned activities. They would also benefit from coaching on Do's and Don'ts of advocacy. Discipline is important for everyone who needs to deliver consistent, clear messages that are intended for the public and policymakers.

The "Cautious" group is different from the other two groups. The idea that Americans are harmed today throughout the United States and that everyone is at risk will probably be a surprise to the people in this group, who comprise nearly a fifth of the population, just as it will be a surprise to the rest of the segments: the "Disengaged," the "Doubtful," and the "Dismissive." These four groups will not understand that some people are at greater risk than others, namely children, pregnant women and their pregnancies, people with chronic health conditions, the elderly, the poor, and many communities of color. They will probably not understand that climate change solutions like clean, renewable energy will lead to cleaner air and cleaner water and thus better health for everyone. As a result, the messaging described in this Primer is important for these groups of people and especially the "Cautious."

Global warming is not generally on the radar of the "Disengaged" group. They are a small percentage of the population and may be focused on their own concerns. The "Doubtful" and the "Dismissive" together comprise almost 20% of the population. While they doubt or reject the idea of global warming, they do have a positive reaction when they learn about the health benefits that accompany global warming solutions, specifically that clean, renewable energy for electric power and other clean energy solutions like electric cars or public transportation will make the air and the water cleaner and lead to better health for everyone.

Framing the Problem

Climate change is already impacting human health; however, the American public doesn't really understand the relationship between climate and health.[16] Research has documented that most Americans do not perceive global warming as a personal threat. As cognitive research has determined, the way that the issue is organized in the human mind and/or discussed can dramatically affect how the problem and its cause are perceived and what, if anything, people understand needs to be done to solve the problem.[17] Until very recently, the way that climate change was presented in the media actually distanced it from people. The main focus was on glaciers, polar bears, plants, and penguins. This nature-oriented idea created the perception that the problem doesn't impact people's lives. Furthermore, the media emphasized the most dramatic impacts, which were not expected until decades into the future, focusing on what was expected by the year 2100. This gave the impression that only the natural world was impacted and that there would not be any significant human effects until the distant future, if ever. Since there was little focus on specific people who were affected, people assumed that there would be no impact, certainly not in the United States to people or their communities. Of course, that is not the case. As explained in Chapter 2, global warming greatly impacts people's health in the United States and around the world.

Additionally, the media also focuses on the economic implications of climate change, threats to national security, the moral obligation of people because of how climate change affects them and the next generation, and the political disputes that have come up. These frames are known as the *environmental frame,* the *economic frame,* the *national security frame,* the *moral frame,* and the *political frame,* respectively. However, a frame that has come up only recently is the *health frame*, which is the frame that explains the relevance of global warming by addressing its impact on personal health.

Frames are very influential in shaping public understanding when the message (which comprises the frame itself) and the messenger are in agreement. The most effective messengers are trusted authorities on the frame being presented. For example, the national security implications of climate change should be explained by national security experts, not health professionals; while the health frame would be most appropriately explained by health professionals. People are more likely to accept information if they believe that the source of the information is trustworthy. However, different people trust different information sources. Health professionals are credible when discussing health topics, and as research has proven, they are credible specifically when discussing the health effects of climate change.[18,19] As a health professional, your credibility will be greater than other people's when discussing the health harms and benefits of climate change and its solutions.

The Health Frame

Most people are unaware of health implications of climate change. The work of the *Climate Change in the American Mind* project shows that health is not top-of-mind when people are asked to share their thoughts about climate change. And yet, the public's personal concerns for their own health and welfare and that of their family are top priorities. Our Declaration of Independence confirms and supports this. The phrase "life, liberty, and the pursuit of happiness" is why we live in an independent democracy. Another reason is that "providing for the general welfare" is justification for establishing the Constitution. When people are specifically asked about the potential consequences of climate change, 75% of respondents say that health is an "extremely" or "very important" concern associated with climate change.[20] Covid-19 added to public concern about health. As public health professionals, we understand that concerns for health are close to our nation's founding principles; as would be expected, this is a concern of all public health professionals. There are many solid reasons to inform and educate the public about the health risks and health harms attributable to climate change and to empower people to work with their health providers to protect themselves and their communities. For many people, when they understand what is at stake, preventing further damage to the climate becomes a compelling reason for action.

To further support this thinking, research has proven that the **Health Frame** is a highly effective way to discuss global warming. Research by Maibach and colleagues has determined that the health frame helps Americans relate to and form an appropriate response to climate change.[21] The political frame causes unnecessary negative issues that undermine agreement and have prevented people from establishing common ground that creates solutions. The health frame will probably not be undermined in the same way. The health frame is also more effective than the environmental or national security frames.[20,22] Another benefit of the health frame is that it catches the interest of more people who would otherwise be less interested in climate change.

Climate change solutions are also pathways to better health. Clean, renewable energy and energy efficiency cause less air pollution. More public transportation and other active modes of transport lead to more walking and less vehicle emissions. Reduced air pollution from reduced burning of fuel for energy and for powering vehicles reduces the production of greenhouse gases and means a more positive impact on the atmosphere and thus on the climate. Because solutions to climate change lead to better health, messaging can focus on both concepts—health and climate—simultaneously.

The concept of "eat and move for wellness and sustainability" merges concepts that fit together easily and address two familiar health topics, healthy eating and physical activity. A diet that is low in meat and high in fruits and vegetables has less impact on the environment and contributes to better health. Meat production, especially raising beef cows, is responsible for the majority of 10% of greenhouse gases that are generated in the United States from the agricultural sector. Those greenhouse gases are mainly from the methane that is produced by cows because of how their digestive systems work. A diet lower in meat is consistent with the Department of Agriculture's current dietary recommendations and could lead to fewer beef cattle and less release of a powerful greenhouse gas into the atmosphere.

Organic agriculture plows last year's plants back into the earth, which **sequesters** (captures) the carbon in those plants in the ground. As mentioned earlier in this Primer, soil is a carbon sink and could sequester and thus **draw down** much more carbon dioxide from the atmosphere than is currently the case. In contrast, the fertilizers used in nonorganic agriculture are produced using fossil fuels. To process nitrogen into artificial fertilizers, natural gas (methane) is needed. Methane is a powerful greenhouse gas that evaporates into the air from drilling sites. The nitrogen fertilizer also evaporates into the air as nitrous oxide, which is another powerful greenhouse gas. Using these fertilizers adds to greenhouse gases. In contrast, plowing last year's plants back into the fields accomplishes similar fertilizing benefit minus the gases that add to global warming.

Walking, biking, and selecting public transportation are better for health because people move physically while traveling. People who take public transportation get more exercise than those who drive.[23,24] Fewer vehicles on the road, especially those that are powered by internal combustion engines—as opposed to electricity—mean less harmful air pollution is emitted into the air that everyone breathes.

Individuals in all segments of Global Warming's Six Americas buy into the concept of "**co-benefits**" or "**benefits**," which is associated with taking action to limit global warming.[25] In other words, messaging about the health benefits of protecting the climate is positive messaging throughout the population represented by the Six Americas, from "Alarmed" to "Dismissive."

Most people across population segments endorse the following statement:

> *"Taking actions to limit global warming—by making our energy sources cleaner and our cars and appliances more efficient, by making our cities and towns friendly to trains, buses, and bikers and walkers, and by improving the quality and safety of our food—will improve the health of almost every American."*[19]

Considering the effectiveness of the health frame and the need for agreement between the frame and the messenger, it makes sense that health professionals are ideal messengers about actions that are beneficial to health and the climate at the same time.[26]

Messages that stress health and suggest behavior that is beneficial to the climate include:

- The healthiest diets for our bodies and for our climate: less meat and more fish, vegetables, and grains. Organically grown food sequesters more carbon from the atmosphere and puts it back into the ground.
- Increasing exercise by walking, biking, or taking public transportation are all active forms of transportation and are better for the heart and leave the air cleaner.
- Cleaner cars and more public transportation, such as buses and trains, leave the air cleaner and encourage people to move more.
- Efficient Energy Star appliances and using energy efficiently through better insulation or by switching off lights and appliances when not in use means less fossil fuel is burned and less pollution will end up in the air and in people's lungs.
- Choosing clean, renewable energy sources creates cleaner air and water and helps everyone live a healthier, more sustainable life.

Two occasions on which automobile traffic was dramatically reduced in American cities created measurable benefits for air quality. In one of those cities, health benefits were clear, specifically a reduced asthma burden.

Case Study 3.1 A reduction in asthma cases at local hospitals was noted during the Olympic games in Atlanta in 1996 when vehicles were restricted and excluded from the Olympic village and the surrounding area located in the heart of downtown. Air quality improved and there was a dramatic, immediate fall in emergency-room visits for asthma patients in surrounding hospitals.[22]

Case Study 3.2 Improvements in air quality were measured in Los Angeles in 2011 when Freeway 405, a major artery, was closed for construction. The closure caused a major reduction in air pollution, which is a known cause of asthma. The reductions were documented at the neighborhood and regional levels.[25,27]

While small changes on the personal level may not seem significant, at a regional or citywide level, they can cause major changes in our exposure now and in the future to the adverse conditions, such as air pollution that affect our health directly, and contribute to the health damages associated with climate change.[28] Diet, active transportation, energy efficiency and Energy Star appliances, and sourcing electricity from clean, renewable energy can all contribute to addressing climate change. Because each of these solutions helps achieve better health—and addresses climate change–people should be receptive when the information is delivered by a health professional. Health professionals have a key role in reaching individuals—and the population in general—regarding climate change.

Public Health Communication Competencies: Inform, Educate, Empower
Inform and Educate

As a *health professional*, the public will trust you more than other messengers as a source about the *health threats* of global warming. You can tell people that their health is at risk with a simple statement; however, an explanation is required to help people understand what that means. The immediate health threats come from direct exposures that most people understand, such as extreme heat lasting for several days, torrential rains and severe storms creating flooding and blackouts, wildfires, heavy pollen seasons, poor air quality, mosquitos and ticks, and contaminated food and water. However, people may not understand that the health consequences are from human caused alterations in the environment. Stories about real experiences are effective because they convey exactly why an experience or environmental condition is a threat to health, explain why its root cause is climate change, and explain why certain groups are at greater risk than others. For example, illustrative stories include one about a child who experienced heat illness after playing outside on a hot day, or one about an adult with asthma who required care from the emergency room when exposed to wildfire smoke, or a family that was displaced and temporarily homeless because of extreme weather and then faced family conflict or substance abuse. These stories illustrate the nature of the problem and who may be at risk. Connecting them with the oncoming changes in the climate requires an explanation that "connects the dots." Otherwise, people may not understand the connection. These conditions are happening more frequently because of climate change.

Because different exposures occur in different communities in different parts of the country, it is better to customize your messaging so that it emphasizes local conditions and dangers. People understand threats more easily when they resonate with them because they share the experience. Customizing messages and stories for specific communities with specific exposures will be most effective. Although everyone is at risk, those who are most at risk include children, pregnant women and their pregnancies (because of greater sensitivity to heat and air pollution), the elderly (due to greater heat vulnerability), people with chronic health conditions (cardiorespiratory, kidney, and mental health conditions), and people with a lower economic status, including many people of color. Messaging should inform everyone, but especially those who are more vulnerable. Specific messaging can help reach those who are vulnerable so they will protect themselves from the dangers that they face.

Empower

Empowering people so they encourage progress toward solutions requires additional information and resources. When public health professionals inform and educate people about their risk, they attempt to reduce the risk by telling them what they can do to protect themselves. Instructions and resources empower people to care for themselves and for their communities. Many communities have already established adaptation resources that reduce risk or protect people. Cooling centers are available in some cities during heat waves. Shelters are available in many communities when extreme weather threatens. Evacuations occur in the event of wildfires. Being prepared is essential in the event of power outages and floods. Preparedness also means having adequate water and food supplies and ongoing sources of information available from battery-powered radios or other devices. Public health professionals and emergency personnel often work together to make communities more resilient against threats. Preparedness is an essential part of public health practice. Equity principles must be an integral part of preparedness planning. Under-resourced and under-represented communities suffer the most in the event of public emergencies. They may require specific resources to meet their needs. For these communities, being prepared and getting what they need are essential. This requires involvement of community members in the planning process.

The Centers for Disease Control and Prevention (CDC) has a Climate Ready States and Cities Initiative, which has helped 16 states and two cities prepare for climate change.[29] The framework for this approach involves identifying and preparing for the specific vulnerabilities in each city or state. That framework is called Building Resilience Against Climate Change Effects (BRACE). The section of the CDC website on these activities has much useful information about the kinds of preparations that are going on around the country.

Empowerment includes the information needed to understand solutions. This, in turn, involves explaining origins and root causes so people understand where to direct their attention. Use of fossil fuels in many aspects of daily life has filled our atmosphere with carbon dioxide and growing amounts of methane, two gases that absorb and re-radiate heat into the atmosphere and expose the population to harmful pollutants in addition to higher temperatures. Remember that the principal gases that comprise our atmosphere, nitrogen and oxygen, do not absorb or re-radiate heat energy. Carbon dioxide, methane, ozone, water vapor, and various other trace gases absorb and re-radiate heat. Hydrofluorocarbons used in refrigeration and air conditioning are also powerful greenhouse gases and still exist in older equipment. To clarify root causes as part of their effort to empower people, the underlying problem—the generation and use of these gases—must be addressed. In contrast, clean energy sources, especially wind and solar energy, do not pollute the air and do not produce waste products that heat the atmosphere.

Changing the power source, fossil fuel vs clean energy, is not controlled by individual persons in all communities. For this reason, policy change at the municipal, county, state, or federal level is necessary. In 2016, The Clean Power Plan, a policy of the Environmental Protection Agency, was finalized. The policy required states to change their energy mix and provided incentives for moving toward clean, renewable energy. This policy was discontinued in 2018. This left states to pursue such policies on their own if they chose to do so. About one-third of the states have done this.

Energy efficiency also offers the opportunity to address the problem. The most efficient use of energy can reduce reliance on the fossil fuels that are polluting the air and causing the climate's temperature to rise. Because over 70% of the electricity used in urban areas heats and cools buildings, maximizing efficiency with the most modern heating and cooling systems and building insulation can reduce dependence on fossil fuels and the resulting carbon (greenhouse gas) footprint.

However, some residential apartments and individual residents will not have the resources to do this on their own. Public housing programs and state- or county-level programs are increasingly available to help improve efficiency and insulation. Improved insulation and more efficient heating and cooling systems can also reduce dampness, mold, and other unwanted components that impact indoor air quality.

Highly effective insulation, Energy Star appliances, and technologies for heating and air conditioning and lighting systems that reduce energy use can help right now to reduce the use of polluting energy sources. A recent report by the American Council for an Energy-Efficient Economy shows how these approaches have beneficial health effects.[30] Furthermore, a solution to our problems may be easier to share if it involves a *positive* choice that we make *for* energy efficiency, in this case one that is produced by technical ingenuity.

Self-efficacy plays a role in empowerment. Self-efficacy is related to a person's belief that they can or have the ability to do something, such as a behavior or action. Believing that taking personal action makes a difference can bolster individual self-efficacy and collective self-efficacy (involving a group) and motivate people to act. Believing that solutions are possible means that people are more likely to adopt solutions. Communicating despair creates fear and hopelessness, and may lead to delay, stagnation, and poor health outcomes. On the other hand, messages that offer hope and attainable solutions elicit change in a positive direction. The goal is to move affected populations from helplessness to action.

Focusing on solutions helps bring people together, even when they have very different ways of thinking.[27] Vastly expanding the use of clean energy requires a change in individual attitudes as well as policy changes made by public officials who are informed, empowered, and willing to make those shifts. However, these officials need the public's support. If people don't believe that taking action will have a positive impact, a sense of helplessness and avoidance or even denial may occur.[29] Empowering people to protect themselves and informing them that they are capable of doing that, creates empowerment that helps people to respond to climate change. Most people are willing and able to protect themselves, especially when they know how to do it effectively—as long as it does not require too much effort or expense.

The general public is already supportive of solutions to global warming. Recent surveys show that throughout the United States, more than 60% of people believe that carbon dioxide (CO_2) should be regulated as a pollutant and more than 50% favor requiring utilities to generate at least 20% of the electricity they use from clean, renewable sources.[31] A larger percentage support more research on clean energy and favor government tax rebates for energy-efficient vehicles or solar panels as an incentive.[32] Yale Climate Opinion Maps are available online and updated every few years so that opinions on these issues may be viewed at the level of the state, county, or congressional district. Respondents in many areas of the country favor a carbon tax on fossil fuel producers.

Of course, purchasing clean, renewable energy or making it possible to do so, is a direct way to participate in the solution. In many parts of the United States, individuals can already purchase clean, renewable energy from electric power producers in their community.[33] Some people are making their own renewable energy by attaching solar panels to their homes. The solutions mentioned above can make a positive impact on our future; people can make a difference.

Creating Targeted Messaging for Action

When people feel engaged in the problem of global warming and are ready to become actively involved, they will need to communicate with other people. They should include key evidence-based messages about climate change and incorporate known strategies for effective communication. Suggesting specific actions that people can take is also important. The value of taking action is that it prevents people from feeling helpless. **One of the best-known truths of communication theory is to use** *"simple clear messages, repeated often, by a variety of trusted voices."* Communication experts frequently share this guiding principle for how to communicate effectively. The principle is broken down in this way:

1. **Use Simple Clear Messages:**[34]
   - The less you say, the more you are heard. Avoid unnecessary, long explanations. Focus on your most important concept. Evidence-based messages, i.e., those demonstrated through research to be understood and have the desired effect are most effective.
   - State the concepts that are the most valuable (in achieving your communication objective). Stick to the basics even if it is difficult. Extraneous facts are not helpful; they may confuse your audience. Evidence is also essential here.
   - Audience research is the most reliable way to determine which messages are the most valuable. This is research with the audience that is the target of the communication. Since different audiences will require somewhat different messaging, evidence-based communication serves you best.

2. **Repeated Often:**
   - Repetition is the mother of all learning (and liking and trust).
   - You may adapt it, elaborate on it (a bit), but no matter what, find ways to state your message early and often.
   - When possible, reinforce (i.e., repeat) the messages with visual images, verbal images (i.e., metaphors), and illustrative stories.
   - According to an often-quoted communication consultant:

     "There is a simple rule, you say it again, and you say it again, and you say it again, and you say it again, and you say it again, and then again and again and again and again, and about the time that you're absolutely sick of saying it is about the time that your target audience has heard it for the first time."

3. **By a Variety of Trusted Voices:**
   - You—as a health professional—are among the most trusted voices in America.
   - You are the most trusted voices on climate and health.
   - If your messages are simple and clear enough, other trusted voices— even members of your target audience—will start repeating them to their friends and family, to their coworkers, and to others. Make that your goal.

General Advice for Health Professionals

Climate change or global warming is real and it is caused by humans. It is harming our health here and now and harming some people more than others. These harms will get worse if we don't do something to stop them, but we know we can address climate change (or global warming) and that this will rapidly improve our health.

- Start with Health: We know that the **Health Frame** is a highly effective way to discuss global warming. It helps Americans relate to and form an appropriate response to climate change. Research also confirms that medical professionals are trusted messengers. People expect you to speak about health, so start there and bring climate change into the discussion.
- Stick to Your Expertise: Don't feel like you must speak about public policy or details about climate science. Confirm that more than 97% of *climate scientists* agree that human-caused climate change is happening and then discuss what you know best: the health impacts.
- Use Stories about People: Be ready with examples from your community or your patients that make the harms associated with climate change a reality.
- Be Solution-Oriented: Always communicate that there is hope—and that clean, renewable energy can improve health today and protect the climate and our health in the future.

Talking Points for Health Professionals

- **In communities across the United States, climate change is harming our health now.** Health professionals understand this because the health of the communities we serve is being harmed in front of us. The rising heat of climate change is increasing heat-related illnesses, worsening allergies and chronic heart and lung conditions like asthma or COPD, and increasing the spread of infectious diseases carried by mosquitoes and ticks. It is intensifying extreme weather events and wildfires that cause injuries and death (use examples that are relevant to your region). The health of any American can be harmed by climate change, but some are at greater risk than others. Children, student athletes, pregnant women, the elderly, people with chronic illnesses and allergies, the poor, and some people of color are more likely to be harmed.

- **There is a scientific consensus that human-caused climate change is happening.** More than 97% of *climate scientists* agree.
- **Unless we work together, these harms to our health are going to get much worse.** As health professionals, we have a responsibility to protect people from further harm. I'm doing that by speaking out to address climate change.
- **The sooner we act, the more harm we can prevent, and the more we can protect and improve the health of all Americans**. The most important thing that we can do to protect our health is to accelerate the transition to clean, renewable energy. This will limit climate change and will rapidly clean our air and water so that we can all enjoy better health. Everyone wants clean air and water and good health. We can accomplish this. Efficient buildings, neighborhoods that are walkable, bikeable, and have public transportation, and smart energy policies are all essential and achievable.

Good communication skills are essential. Like with any public health initiative, cultural competency is critical for effective communication and for achieving desired health outcomes. It is important to assess and understand your audience's beliefs regarding the health issue, as well as the general norms of the culture and society, and to customize messages accordingly. It is also important to understand how and to what degree the community is affected in addition to the projected health outcomes. If you are part of an organized messaging campaign, pre- and post evaluation is critical to accurately anticipating the impact of your messages now and in the future.

Messaging by Climate-Related Issues and Health Risk

Another way to communicate about climate and health as a health professional is to identify the health risk factors that you are most comfortable speaking about and that will resonate with your audience. For example, if you are in an area prone to wildfires or Lyme disease, speak about the risks of those climate-related issues. These facts about health can be ice breakers. You can ask what concerned citizens should do about these problems. Sharing ideas can follow.

- **Extreme heat:** More extreme heat caused by climate change is making more of us ill. More frequent heat waves, greater humidity and extreme temperatures cause more heat illness and are dangerous for children, pregnant women, and people with chronic conditions like lung and heart disease or mental health conditions. People who work outdoors, the elderly, and people on certain medications are also likely to suffer more than others from heat.

- **Air Pollution:** The quality of the air that we breathe is getting worse. Whether it's pollen in the air because of earlier and longer growing seasons, or dangerous particles in the air because of drought or wildfires, or more smog (ozone) because of what happens when heat and light act on fumes, people are more ill because of the changes to our air. It does not just affect people with lung conditions—poor air quality affects the developing lungs of children and causes heart problems in adults.
- **Extreme Weather:** Extreme weather, such as torrential rains, storms, floods, and droughts are happening more frequently and more severely. These events cause injuries, power outages, and make it harder for patients to get health care. In some cases, it can even make water and food supplies unsafe.
- **Water Contamination**: With the changing climate, heavier and more frequent rains are occurring with greater frequency. This can lead to sewage overflows and contamination of crops or waterways. Rains can carry fertilizers into lakes and lead to overgrowth of algae. Some of the algae is toxic and can affect drinking water or lead to restrictions on our recreational lakes. Increasing temperatures are also producing dangerous algae blooms in lakes and coastal areas; and can promote bacterial growth, such as *Salmonella, E. coli, Vibrio*.
- **Ticks and Mosquitos**: Ticks and mosquitos are moving into new areas and multiplying more quickly because of the changing climate. Earlier springs and changing weather patterns lead to diseases that are spread by mosquitoes or ticks in new parts of the country and in more seasons. Whether it's Lyme disease, West Nile virus, dengue, or Zika, people are growing increasingly concerned. It is important to protect yourself and your family from diseases that are carried by these insects.
- **Food Quality and Nutrition:** A changing climate can make our food less nutritious by altering the protein content of wheat, rice, and corn. When these grains are grown in air that has a greater carbon dioxide concentration, the nutrient content of the grain changes and contains less protein.
- **Mental Health:** Extreme weather can separate people from their families and communities, which places them at risk from the negative mental health effects of displacement. Additionally, hotter days can decrease the effectiveness of medicines that people need to take for mental health problems.

Always finish with a hopeful vision that conveys how humans have the ability to positively affect solutions. The threats of global warming are great but the opportunities associated with addressing global warming are also great. We have the opportunity to create a healthier world if we change our energy choices, selecting clean renewables rather than polluting fossil fuels. The result will be clean air and clean water. They will improve the health of the people who breathe the cleaner air and drink the water without toxic substances. Children will be able to develop with healthier lungs and fewer asthma attacks, adults with lung conditions can be hospitalized less, and our population can have a new normal in which communities save some of the money they are spending on medical care.

We can choose a cleaner greener environment, and an economy stimulated by new technologies and jobs. This is not a fantasy; it will be the direct result of relying less on burning fuel for energy and transitioning to new forms of energy production, less polluting forms of transportation, and increased energy efficiency. While we are working to achieve that vision, we can protect people from health harms. The people with whom we share protective information can help themselves and others. That knowledge will be valuable and universal.

Frequently Asked Questions

Note: These questions use "health professional" but they are applicable to any healthcare worker or discipline.

- **Question:** "You're a health professional. How much do you really know about the science of climate change?"

   Possible response: "I know that climate change is real and caused by humans because more than 97% of *climate scientists* agree on this. And I see how it is affecting the health of the people in our community."

- **Question:** "As a health professional, shouldn't you be more concerned with health reform or healthcare issues?" (Alternatively, the question could be about focusing on insurers or some other health policy issue).

   Possible response: "As a health professional, my priority is the health of the population I serve. I'm speaking out and advocating because it is fundamental to how well I can work to improve and protect the health of my community. Not all medicine is about treating illness once it has occurred. This is really preventive care."

- **Question:** "Isn't this just a partisan attack or politically charged rhetoric?"

   Possible response: This is not about politics. Health has nothing to do with the political views of individuals, communities, or healthcare providers. We were all vulnerable to Covid-19 and we could all suffer the health impacts of climate change.

- **Question:** "How can you attribute asthma increases (or other health harm) to climate change?"

   Possible response: The research confirms that climate change is increasing the frequency and intensity of many of these threats to our health. None of these risks are new, but climate change is changing how it impacts Americans and the severity of the impact. Allergies are a great example. They affect many people, but now, due to global warming, the pollen season is longer and stronger. People start having symptoms earlier in the year and need more medicine to treat them. Wildfires are another good example of how this works because they are increasing as a result of climate change. During a wildfire, smoke travels hundreds of miles and is inhaled by people who are adversely affected by the smoke. This leads to asthma attacks, emergency visits, and hospitalizations for those with any underlying lung condition. People with heart conditions are also affected.

Five Evidence-Based Beliefs that Predict Appropriate Action

Communication research has identified five key beliefs about climate change that predict whether a person will take appropriate action in the form of consumer or political activism or support for a societal response. Here are the five beliefs

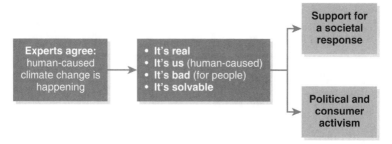

Figure 3-3 Five key beliefs about climate change and appropriate attitudes and actions.
Ding-Ding et al. 2012; Lewandowsky et al. 2012; Rose-Renouf et al. 2014; Krosnick et al. 2006.

stated very briefly: 1) climate change is real, 2) it is caused by humans, 3) it is harmful to people, 4) experts agree that it is real and caused by humans, 5) it is solvable. **Figure 3-3** shows that accepting the fact that humans caused climate change is essential.[35]

These messages, which relate to each of the five evidence-based beliefs, can be delivered in a way that emphasizes health.[36]

1. There is a scientific consensus about human-caused climate change.
2. In communities across America, climate change is harming our health now.
3. The health of any American can be harmed by climate change, but some of us are more vulnerable and face greater risk than others.
4. Unless we work together, the harms to health are going to get much worse.
5. The most important action we can take to protect our health is to reduce heat-trapping pollution by reducing energy waste and accelerating the inevitable transition to clean, renewable energy. This will address climate change and give us cleaner air and water.

These messages can be formatted into a message box to make them easier to remember. See **Figure 3-4**. However, you may wonder how you should communicate these messages. This is easily answered. In short, be the health professional whom you are because it does bestow you with public trust. Also, feel free to express your concerns and hopes as a person. Show appropriate concern and a sense of urgency but remain optimistic. Make it clear that a range of actors should be taking action—citizens, businesses, and government officials. Each of them can take steps to accelerate the inevitable transition to clean energy.

The next chapter will focus on the policy solutions that will address climate change and benefit the public's health. As people become more aware of the health harms and risks to their future, their children's future, and their grandchildren's future, they realize that climate change is a public health emergency, and they become more interested in advancing solutions. As explained in the next chapter, a large group of medical, nursing, and public health associations have come together with hospitals and health systems, and schools of medicine, nursing, and public health to support a policy agenda. This agenda is discussed in Chapter 4.

Problem:
In communities across the nation, climate change is harming our health now. Doctors know this because we're seeing the health of our patients being harmed. These harms include heat-related illness, worsening chronic illness, injuries and deaths from dangerous weather events, infectious diseases spread by mosquitoes and ticks, illness from contaminated food and water, and mental health problems.

So what?
The health of any American can be harmed by climate change, but some of us are at greater risk than others. Children, student athletes, pregnant women, the elderly, people with chronic illnesses and allergies, and the poor and those who have experienced racism or environmental injustice are at greater risk.

Issue: Climate change is bad for our health
More than 97% of climate scientists have concluded that human-caused climate change is happening, and research has proven that it is already harming the health of many of us. As a health professional, I have a duty to protect people from further harm by taking steps to address climate change.

Benefits?
Unless we take concerted action, these harms are going to get much worse. The sooner we take action, the more harm we can prevent, and the more we can protect the health of all Americans. Everybody wants clean air, clean water, and better health.

A useful way to think about it: What's good for our climate is good for our health, and what's good for our health is good for our climate.

Solutions?
The most important action we can take to protect our health is to reduce heat-trapping pollution by reducing energy waste and accelerating the inevitable transition to clean, renewable energy. It is well within our power to accomplish this. Accelerating the transition to clean energy has the added benefit of cleaning up our air and water so that we can rapidly enjoy better health.

Figure 3-4 The Message Box.
Center for Climate Change Communication at George Mason University, 2016.

WRAP-UP

Discussion Questions

1. Of the 10 essential public health services in Figure 3-1, which are the most important for addressing climate change and why?
2. Do you agree that the health frame is likely to help most for influencing how Americans think about climate change? Why or why not?
3. Which two groups in *Global Warming's Six Americas* depicted in Figure 3-2 are the most important to engage and activate and why?

References

1. Huelskamp T, Bast JL. Freedom Rising. The Heartland Institute 2018. https://www.heartland .org/_template-assets/documents/2018_Prospectus.pdf. Accessed September 12, 2019.
2. The Heartland Institute. Freedom Rising. Climate change. 2020. https://www.heartland.org/topics /climate-change. Accessed September 12, 2019.
3. The Public Health System & the 10 Essential Public Health Services. https://www.cdc.gov /publichealthgateway/publichealthservices/essentialhealthservices.html. Accessed July 26, 2020.
4. Centers for Disease Control and Prevention (CDC). The public health system & the 10 essential public health services: the public health system. https://www.cdc.gov/stltpublichealth /publichealthservices/essentialhealthservices.html. 2020. Accessed November 15, 2018.
5. Maibach E, Parrott RL. *Designing Health Messages: Approaches From Communication Theory and Public Health Practice*. Thousand Oaks, CA: Sage Publications; 1995.
6. Health Behavior and Health Education: Theory, Research, and Practice. Editors: Glanz K, Rimer B, Viswanath K. Jossey Bass, 2008.
7. Ibid.
8. Witte K, Meyer G, Martell DP. *Effective Health Risk Messages: A Step-By-Step Guide*. Thousand Oaks, CA: Sage Publications; 2001.
9. Ballew MT, Leiserowitz A, Roser-Renouf, et al. Climate change in the American mind: data, tools, and trends. *Environment: Science and Policy for Sustainable Development*. 2019;61(3):4-18.
10. George Mason University Center for Climate Change Communication. *Climate Change in the American Mind*. 2018. https://www.climatechangecommunication.org/climate-change-in-the -american-mind/. Accessed May 28, 2020.
11. Frumkin H, Hess J, Luber G, Malilay J, McGeehin M. Climate change: the public health response. *Am J Public Health*. 2008;98(3):435-445.
12. Russill C, Nyssa Z. The tipping point trend in climate change communication. *Global Environmental Change*. 2009;19(3):336-344.
13. Boykoff, MT, Boykoff JM. Balance as bias: global warming and the US prestige press. *Global Environmental Change*. 2004;14(2):125-136.
14. Health Behavior: Theory, Research, and Practice. 5th ed. In: Glanz K, Rimer BK, Viswanath K, eds. Jossey-Bass; 2015.
15. Roser-Renouf C, Leiserowitz A, Maibach E, Feinberg G, Rosethal S. *Global warming's six Americas, 2014*. Yale University and George Mason University. New Haven, CT: Yale Project on Climate Change Communication; 2015.
16. Sarfaty M, Maibach E. Communication. Climate change and public health. In: Levy B, Patz J, eds. Oxford University Press; 2015:255-269.
17. McCright AM, Dunlap RE. Anti-reflexivity. *Theory, Culture & Society* 2010;27(2-3):100-133.
18. Roser-Renouf C, Stenhouse N, Rolfe-Redding J, Maibach E, Leiserowitz A. Engaging diverse audiences with climate change: message strategies for global warming's six Americas. In: Hansen A, Cox R, eds. The Routledge Handbook of Environment and Communication. New York: Routledge; 2015.
19. Maibach EW, Kreslake JM, Roser-Renouf C, Rosenthal S, Feinberg G, Leiserowitz AA. Do Americans understand that global warming is harmful to human health? Evidence from a national survey. *Annals of Global Health*. 2015;81(3):396-409.
20. Maibach EW, Nisbet M, Baldwin P, Akerlof K, Diao G. Reframing climate change as a public health issue: an exploratory study of public reactions. *BMC Public Health*. 2010;10(1):299.
21. Hart Research Associates. Messaging recommendations on climate change. Memo to Climate Action Campaign, May 31, 2013.
22. Myers TA, Nisbet MC, Maibach EW, et al. A public health frame arouses hopeful emotions about climate change. *Clim Change*. 2012;113:1105-1112.
23. Freeland AL, Banerjee SN, Dannenberg AL, et al. Walking associated with public transit: moving toward increased physical activity in the United States. *Am J Public Health*. 2013;103(3): 536-542.
24. Rissel C, Curac N, Greenaway M, et al. Physical activity associated with public transport use—a review and modelling of potential benefits. *Int J Environ Res Public Health*. 2012;9(7):2454-2478.

25. Myers TA, Maibach EW, Roser-Renouf C, Akerlof K, Leiserowitz AA. The relationship between personal experience and belief in the reality of global warming. *Nature Climate Change*. 2013;3:343-347.

26. Kennedy B. Most Americans trust the military and scientists to act in the public's interest. *Pew Research Center.* http://www.pewresearch.org/fact-tank/2016/10/18/most-americans-trust-the-military-and-scientists-to-act-in-the-publics-interest/. 2016. Accessed November 17, 2018.

27. Brenan M. Nurses keep healthy lead as most honest, ethical profession. 2017. https://news.gallup.com/poll/224639/nurses-keep-healthy-lead-honest-ethical-profession.aspx?g_source=CATEGORY_SOCIAL_POLICY_ISSUES&g_medium=topic&g_campaign=tiles. Accessed November 17, 2018.

28. Beckage B, Gross LJ, Lacasse K, et al. Linking models of human behaviour and climate alters projected climate change. *Nature Climate Change*. 8(1):79.

29. Friedman MS, Powell KE, Hutwagner L, Graham LM, Teague WG. Impact of changes in transportation and commuting behaviors during the 1996 summer Olympic games in Atlanta on air quality and childhood asthma. *JAMA*. 2001;285(7):897-905.

30. Hong A, Schweitzer L, Yang W, Marr LC. Impact of temporary freeway closure on regional air quality: a lesson from Carmageddon in Los Angeles, United States. *Environ Sci Technol*. 2015;49(5):3211-3218.

31. Choi W, Hu S, He M, et al. Neighborhood-scale air quality impacts of emissions from motor vehicles and aircraft. *Atmospheric Environment*. 2013;80:310-321.

32. Centers for Disease Control and Prevention (CDC). CDC's climate—ready states & cities initiative. https://www.cdc.gov/climateandhealth/climate_ready.htm. Accessed November 8, 2018.

33. Hayes S, Kubes C. Saving energy, saving lives: the health impacts of avoiding power plant pollution with energy efficiency. *Res Report H1801 from the American Council for An Energy Efficient Economy*. Washington, DC.

34. Marlon J, Howe P, Mildenberger M, Leiserowitz A, Wang X. Yale climate opinion maps 2018. *Yale Program on Climate Change Communication*. 2018.

35. Sarfaty M, Gould R, Maibach E, and Burness Communication. Medical Alert! Climate change is harming our health. Report March 2017. Accessed at docsforclimate.org on December 24, 2019.

36. Ibid.

CHAPTER 4

Climate and Health Policy

Every generation leaves behind a legacy. What that legacy will be is determined by the people of that generation. What legacy do you want to leave behind?

John Lewis, American Statesman and
Civil Rights leader, representing the Georgia
5th Congressional District from 1987–2020

KEY TERMS

Black Lung
Cap and Trade
Criteria Pollutants
Draw Down
Fenceline Community
Fracking
IPCC

LEED Certified
Paris Treaty
Silicosis
Redlining
Renewable Energy Portfolio
World Health Organization (WHO)

CHAPTER OBJECTIVES

1. Explain why placing a priority on health is the strongest theme for health professionals who are interested in affecting climate policy.
2. Describe how major mitigation policies are beneficial to health.
3. List the governing tools that governments have to address climate and health policy.
4. Discuss three policy recommendations from the U.S. Call to Action on Climate, Health, and Equity: Policy Action Agenda, and explain how they would benefit population health.

Climate Change and the Social Determinants of Health

For health professionals, addressing the accumulating greenhouse gases in the atmosphere is a priority. Why is it a priority? Because climate change is caused by this accumulation and is already having a profound impact on human health. Food, water, shelter, and safety from harm are essentials for a decent life, and all are threatened by climate change. This makes it an easy decision for health professionals to choose to engage their time and attention on the problem of climate change and work to advance solutions—especially those solutions that we know are specifically beneficial to health.

The warming climate and all the conditions of life that are affected by it have become powerful contributors to the disparities already faced by poor and minority communities. Historical, social, and environmental factors are converging to create the greatest risk for people who live in the most exposed and deprived neighborhoods. The social determinants of health and the interplay between them and environmental exposures create a powerful dynamic, which is overwhelming other factors. An example of a historical factor is offered by new research showing that minority communities, especially African American neighborhoods, facing **redlining** in the past are now the hottest parts of many cities. In other words, redlining from banks, which made it impossible for inhabitants to secure the loans they needed to purchase their own homes, determined a deprived future for those who missed out on the loans.[1] They and their families remained on neglected streets laden with concrete apartment complexes and little greenspace or tree canopy coverage where the urban heat island effect means day and nighttime temperatures are 10–15°F higher than elsewhere in the area. Many of these communities have the worst housing in parts of town that are also more likely to flood—an environmental exposure also frequently shaped by neglect of these communities. The residents live on streets that abut major roadways where proximity to vehicle traffic, another environmental risk factor, means more polluted air and greater likelihood of asthma and poor lung development in childhood.[2] Poor air quality is associated with greater risk of mortality during the Covid-19 pandemic.[3] Neglect of housing upgrades by authorities or owners who are distant from the local streets is a social factor that often means that efficiency of energy use in the apartments in those neighborhoods is poor so that warmed or cooled air escapes and fails to heat or cool as intended. This can drive up heating and cooling costs so they are too large a share of monthly income.[4] Many poor and minority neighborhoods are locations that pose an unhealthy trap for their residents.

This reality means that close attention to high-risk neighborhoods is needed in order to prepare and protect the U.S. residents who live there from the health harms of climate change. Local officials must be made aware of these realities. Attention to these problems can become an imperative to act ethically by making equity a prominent part of efforts to reduce the health threats from climate change. Residents of higher risk neighborhoods should be engaged in identifying and planning solutions. Health professionals can play a key role in drawing attention to these problems, validating the concerns of local residents, and supporting principles of community participation when addressing them.

This chapter will offer a very brief review of policy solutions to addressing climate change and improving health. Despite the pressing importance of adaptation solutions to reduce pressing risks, most of the solutions in this chapter are mitigation solutions—reducing the contribution of greenhouse gases that are still accumulating and thus preventing further damage to the climate—as opposed to adaptation solutions that protect people from the current and anticipated health impacts from climate change. Some mitigation solutions have adaptation concerns woven through them, and this is highlighted in the text. Legislative and regulatory solutions are included here; many are policies that can be implemented at the national or state level.

Sources of Accumulating Greenhouse Gases and How to Mitigate Them

The largest contributors within the United States to the rapidly accumulating concentration of greenhouse gases in the atmosphere are generation of electricity (from coal, oil, and natural gas), transportation (powered by gasoline), industrial processes (powered by fossil fuels), and agriculture. See the **Figure 4-1** below for the relative share of greenhouse gases that are contributing to global warming by different categories of economic activity and for the sources of these gases. The last item, agriculture, contributes to greenhouse gas accumulation for two reasons: 1) because raising cattle results in methane production, and 2) fertilizers are generated in a process that involves two greenhouse gases: methane gas is involved in producing fertilizer, and the nitrogen in the fertilizer combines with oxygen in the air to form nitrogen dioxide—another powerful greenhouse gas.

The health sector is not on this graph. However, the combined energy intensity and resource use of the health sector accounts for about 10% of greenhouse gas use. While that 10% is contained within the other sections of the pie graph, understanding the contribution of the health sector to climate change can point the way to opportunities to reduce greenhouse gas production.

This chapter includes a policy agenda, which is a roadmap to reducing emissions of greenhouse gases from the generation and use of electricity, and from transportation, agriculture, and health care. As mentioned earlier, use of energy in buildings for heating and cooling and electricity accounts for a significant percentage of all energy use. Buildings in the United States use over 70% of all electricity, and account for 40% of all energy uptake. This is an essential target for efforts to reduce energy use, which can be approached with a focus on energy efficiency.

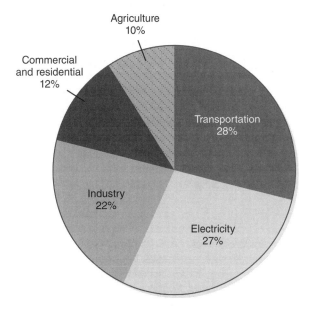

Figure 4-1 Total U.S. Greenhouse Gas Emissions by Economic Sector in 2018.

Approaches to Mitigating Global Warming

If the problem of global warming is defined as greenhouse gases (heat trapping pollution such as carbon dioxide, methane, nitrous oxide, hydrofluorocarbons) accumulating in the atmosphere, the solution should focus on the sources of those emissions and how changing policies and practices can positively affect the situation. The goal is to minimize the contributions to greenhouse gas accumulation and transition to practices that do not emit greenhouse gases. This will reduce the damage to the climate and directly benefit human health.

Solving climate change will benefit the health of the U.S. population and the entire world. It will go a long way to contributing to equity and environmental justice, most immediately for "fenceline" communities. **Fenceline communities** are close to the fossil fuel extraction, refining, or storage process and are thus exposed directly to harmful gaseous emissions. Drilling or **fracking**, may seem like the right approach to producers, but cause environmental risks to communities that are located near the sites.[5]

Sources of Greenhouse Gases

The pie graph above shows the U.S. sources of CO_2 that continue to enter the Earth's atmosphere. The main sources are transportation powered by fossil fuels (on the ground and in the air) (28%), electricity generated in the energy sector by fossil fuels (coal, natural gas, petroleum) (27%), agriculture (especially the cattle-raising component) (10%), industry (22%), and commercial and residential power (12%). The health sector of the economy also accounts for 10%. If this is the starting point, solutions should focus on decreasing the output from all of those sources. It should be noted that since the relative percentage of emissions by sector differs by country, the global percentages will also differ from what is shown in the Figure 4-1.

Transportation

The transportation sector, which is the largest contributor to greenhouse gas accumulation in the United States, is mostly powered by fossil fuels. However, this could change dramatically if more cars, trucks, trains, and trolleys run on renewable energy sources (hybrid and electric vehicles) and if there were many more opportunities for "active transportation" through walking, biking, and electric scooters, etc. The population's health would benefit if carbon emissions were reduced from the transportation sector because people would no longer be exposed to the harmful pollutants that are damaging to cardiorespiratory health. Being more active during transportation would also benefit personal health.

Electricity

Generating electricity in the energy sector still relies mostly on fossil fuels, but clean, renewable energy sources such as solar (photo voltaic cells) and wind (turbines) are increasing rapidly and have become economically feasible in many energy markets. In 2018, clean, renewable energy provided 36% of U.S. electricity production (19% from nuclear, and 16% from wind, solar, and hydropower), while fossil fuels provided 64%.[6] In specific states like Iowa, where 40% of the electricity comes from wind turbines, clean, renewable energy provided more.

Coal Burning

Coal use has decreased markedly in the United States but in many areas of the world, coal use is increasing. This continues to be a significant health issue for two reasons. One is that coal is the "dirtiest" source of electric energy because more particulate matter or soot is produced when it is burned compared to other fuels. More carbon dioxide is also produced than from other fossil fuels. The particulates are associated with lung and heart disease and stroke. The smoke contributes to air pollution immediately as it leaves smokestacks, and it presents ongoing health dangers long afterward because poisonous metals in the coal like mercury, lead, and cadmium spread with the smoke over large distances, settling into rivers and streams. Fish and the people who catch and eat them eat those poisons at the same time.

Coal emissions are extremely toxic pollutants. The coal ash that is left after the coal is burned can be even more hazardous because it is left on the ground for long time periods and can leech into the groundwater used for drinking water. Occasionally, there is a disastrous overflow of this poisonous coal ash into local waterways.[7] Coal mining also exposes miners to coal dust that can cause "**Black Lung**" disease, a debilitating chronic lung disease. Recently, coal is being extracted from mines that contain more rock and less coal. This exposes the miners to dust that causes **silicosis**, another disabling chronic lung condition.

Oil and Natural Gas

Exposure to emissions from extracting and refining the other principal fossil fuels, oil and natural gas, also damage health. In recent years, evidence has emerged that shows that exposure to drilling sites has negative impacts on developing infant brains, likelihood of a poor outcome in pregnancy, and declining neurocognitive function in the elderly.[8-10]

A study published in December of 2017 in Pennsylvania found a higher incidence of premature and low birthweight babies born to mothers who lived close to fracking operations than for those who lived farther away. Infants born within .6 miles (1 kilometer) of a well were 25% more likely to have low birth weight than those born 1.8 miles (3 kilometers) away. They also obtained lower scores on standard indices of infant health. Babies who were born between the two distances had intermediate outcomes.[11] Evidence exists that proximity to coal and oil drilling and refining is also an issue.[12] In areas in which coal and oil plants have closed, preterm births have decreased. When California closed eight coal or oil-fueled electric plants, a significant decline occurred in preterm births among women living nearby. Preterm births decreased from 7 to 5%. These issues have been identified fairly recently and more research is required going forward.[13]

People's health will benefit directly if policies that protect the affected populations—especially those who live nearby—from activity such as drilling are implemented. There will also be a significant health benefit from decreasing the damage to the climate.

Industry is another source of greenhouse gases, with some industries, such as the production of cement, producing more greenhouse gases than others. A discussion of industrial sources of greenhouse gases is outside the scope of this Primer.

Buildings

Commercial and residential buildings use electric power for heating, cooling, lighting, and appliances. Although most of that power is still generated by fossil fuels, an improved energy-saving plan and efficiency can have a substantial impact on the amount of power used. The less power that is used, the less energy must be extracted, generated, and burned. For example, in urban areas, buildings account for over 70% of electricity use (across all regions, buildings account for 40% of energy use). If the energy efficiency would be maximized in buildings, energy demand would drop, and with it, the need for and burning of fossil fuels and the human exposure.

Several components determine energy use in buildings. Insulation ensures less energy loss associated with heating and cooling. A better design allows daylight rather than artificial lighting to be used whenever possible and decreases the demand for electricity. Installation of state-of-the-art appliances, including those that adjust to the occupants' needs and shut down or standby automatically when not being used can reduce energy use significantly. Examples include movement-activated light switches, timed thermostats, maximally efficient heating, and air conditioning units. These innovations can assure significant reductions in power use. If all the "off the shelf" technologies were used, we could decrease power use by 25–40% along with the fossil fuel combustion needed to produce that power. Scientists have also estimated the health benefits and associated monetary savings that accompany maximizing energy efficiency.[14] Many of these efficiency features are specified in the standards of the U.S. Green Building Council and are incentivized or required by local ordinances.[15,16] **LEED Certification** through the Council establishes the presence of a suite of these features.

Health Care

The contributors to emissions can be more closely examined by looking at the economic sectors. The health sector as a whole is extremely energy intensive. It is responsible for approximately 10% of greenhouse gases that enter the Earth's atmosphere.[17,18] Many people who work in the health sector are concerned that their work is part of the problem. The organizations Health Care Without Harm and Practice Green Health have made efforts to eliminate the adverse impacts of the health sector on the environment. This means decreasing waste, and energy use, and increasing energy efficiency.[19]

Agriculture

Another major contributor to the greenhouse gases that enter the atmosphere is the agricultural sector. In the United States, this accounts for about 10% of greenhouse gas output. It is 25% worldwide. A large percentage of this comes from raising meat for consumption. Raising cattle, directly adds methane—a greenhouse gas more potent than carbon dioxide—to the atmosphere. Reducing beef consumption can help reduce the damage to the climate. Low meat-intake diets, on a macro level, could reduce conversion of forest lands to grazing lands, leaving more trees to absorb carbon dioxide and act as a carbon sink. Diets lower in meat that include more grains and vegetable protein can also positively contribute to personal health.[20] See the relative greenhouse gases emitted into the atmosphere and their sources in **Figure 4-2**.

More grazing land is often created at the expense of land that was previously forested. Wild forested land, such as the Brazilian rain forest or the rain forests of

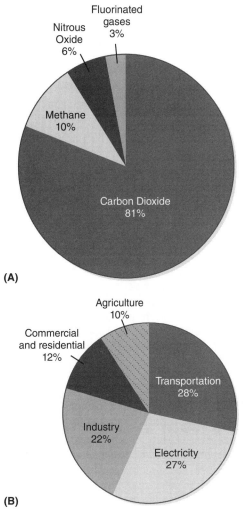

Figure 4-2 (A) Overview of Greenhouse Gas Emissions in 2016; **(B)** Sources of Greenhouse Gas Emissions in 2016.

Reproduced from United States Environmental Protection Agency. Sources of Greenhouse Gas Emissions. https://www.epa.gov/ghgemissions/inventory-us-greenhouse-gas-emissions-and-sinks.

Indonesia, offers carbon sinks for the increased carbon in the atmosphere. Grazing and agriculture can reduce the forest growth available to serve as a carbon sink.

There are other factors to consider. Loss of biodiversity results from cutting down forests for planting and grazing and raising cattle. Raising cattle uses a very large amount of water. In a situation in which water shortages are an increasing problem because of glacial melt and droughts, it is important to use water conservatively and sensibly.[21-23]

The contribution of the sectors shown above to the accumulation of greenhouse gases can be addressed through policies that would reduce the damage to our climate while benefiting our health at the same time. Energy production is one of the two largest sources of greenhouse gas emissions in the United States.

A shift to energy sources that do not pollute and do not produce greenhouse gases is possible if we continue to expand our use of clean, renewable energy and we stop using fossil fuel, starting with coal, which produces the most CO_2. Energy efficiency can offer an equal capacity to power our lives while relying on smaller amounts of pollution-causing fuels. Low or no carbon transportation will leave the air cleaner and cause fewer lung health problems, while effectively transporting people and promoting economic growth. The health sector can significantly reduce its wasteful energy and resource use. Better land use and agricultural methods that sow last year's carbon-rich plants back into the soil rather than rely on nitrogen containing fertilizers produced in a proces that relies on methane can optimize the use of agricultural soil as a carbon sink. Composting food waste can produce soil instead of contributing to methane production that comes from rotting food in landfills.

The Call to Action on Climate, Health, and Equity: A Policy Action Agenda (Agenda) below, draws on these approaches. It also addresses the displacement that impacts communities either as a result of damaging floods and storms or economic transitions that cause some industries to thrive and others to shrink. Some people and communities will be affected by economic changes; some by the direct negative impacts of climate change. Much of the Agenda directly benefits health, but the document starts by declaring climate change a public health emergency. Climate change is a health emergency because millions of human lives are threatened by the conditions it creates, in the form of searing temperature, torrential rain, immense fires, powerful storms, and dangerous new infectious diseases, and in the way it threatens our shelter, sources of food, and the communities in which we live. It threatens our health and everything we rely on to survive.

The Call to Action of the Medical and Public Health Communities

In 2019, *A Call to Action on Climate, Health, and Equity: A Policy Action Agenda* was announced by more than 70 health organizations. This announcement received significant attention from the press with coverage by more than 470 media outlets.[24] The coverage focused mostly on the declaration of a public health emergency by so many prominent mainstream medical and health organizations including the American Medical Association, the American Nurses Association, the American Academy of Pediatrics, and the American Academy of Family Physicians. The actual policy recommendations in the agenda addressed electric energy generation and use, energy efficiency, transportation energy use and emissions, agricultural techniques and use of lands, and reduction of energy use by sectors such as health care. The document defines an agenda of climate change solutions that is well researched and has been endorsed by more than 150 major medical, nursing, and public health organizations, including a group representing 500 hospitals and health systems, and many health education institutions.[25]

The source of this policy agenda is the lead up to a *Climate Action Summit* (September 2018) called by the sitting Governor of California Jerry Brown and the former Mayor of New York City Michael Bloomberg.[26] That event brought states and municipalities together from across the United States to discuss their efforts to address climate change. The 2018 Call to Action was delivered to the 24th United Nations meeting on climate change in Poland (referred to as the 24th Convening of the Parties or "COP 24"), which reunited 195 countries that previously agreed to the Paris Treaty on Climate Change. Fundamental to the concept of the initial version of the agenda is emphasis on equity and the recognition that communities that have already suffered environmental injustice are in the greatest danger from the growing threats and health impacts of climate change. The **World Health Organization (WHO)** drew on this agenda for recommendations in its publication, "COP24 Special report: Health & Climate Change" that was published at the time of the COP 24.[27]

The *Paris Treaty* of 2015 produced an agreement to take action to limit the warming of Earth's atmosphere to 2 degrees Celsius with best efforts to limit the warming to 1.5 degrees Celsius. Each country defined its own contribution, called a nationally defined contribution or NDC to this joint solution. Some funding was established to help developing countries that have contributed the least to the problem to help them incorporate clean energy into their development plans. The United States was heavily involved in negotiating this treaty and supported it. However, in 2017, six months after President Donald Trump was elected, he announced that the United States would abandon the Paris Treaty. Despite this announcement, 20 states and 50 cities announced that they would continue to abide by the treaty. Today, there are 247 cities that have announced their intent to abide by the treaty. They use the slogan, We are Still In.

The *U.S. Call to Action on Climate, Health, and Equity: A Policy Action Agenda*, which came out of the Climate Action Summit and the Paris Treaty, may be found with all of the current endorsers at its own website (climatehealthaction.org). It is reprinted below under the heading 'A Call to Action on Climate, Health, and Equity.

A Call to Action on Climate, Health, and Equity

Climate change is one of the greatest threats to health America has ever faced—it is a true public health emergency. The health, safety and wellbeing of millions of people in the United States have already been harmed by human-caused climate change, and health risks in the future are dire without urgent action to fight climate change. As former Surgeons General Richard Carmona and David Satcher said: "We're all at risk and our leaders must lead on global warming. Now."[28] But the health crisis caused by climate change also presents a major health opportunity. Building healthy energy, transportation, land use, and agriculture systems will now deliver immediate and sustained health benefits to all and reduce future health risks from climate change.

Our organizations represent physicians; nurses; health and public health professionals and health workers; hospitals and healthcare systems; health education institutions; and public, environmental, mental, and community-based health agencies and organizations. We have dedicated our lives to improving the health of our patients and communities.

Therefore, we call on government, business, and civil society leaders, elected officials, and candidates for office to recognize climate change as a health emergency and to work across government agencies and with communities and businesses to prioritize action on this Climate, Health, and Equity Policy Action Agenda.

Climate change is the "greatest public health challenge of the 21st century." Extreme heat, powerful storms and floods, year-round wildfires, droughts, and other climate-related events have already caused thousands of deaths and displaced tens of thousands of people in the United States from their homes, with significant personal loss and mental health impacts, especially for first responders and children. Air pollution, whose primary driver—fossil fuel combustion—is also the primary driver of climate change, causes hundreds of thousands of deaths in the United States annually. Mosquito- and tick-borne diseases are spreading to new communities. The agricultural, food, and water systems we depend on for our survival are under threat. Without an urgent and effective response, these harms will greatly increase.

Action to reduce climate change can dramatically improve health. Many policies that move us toward safe climate goals have demonstrable and significant health benefits. Climate action in the energy, transportation, land use, housing, agricultural, and other sectors has the potential to avoid thousands of deaths in the United States and millions of deaths each year globally. A just transition to clean, safe, renewable energy and energy efficiency; sustainable food production and diets; active transportation; and green cities will lower climate pollution while simultaneously reducing the incidence of communicable and noncommunicable disease, improving mental health, and promising significant healthcare cost savings.

Equity must be central to climate action. Climate change threatens everyone in the United States; however, it is a more immediate danger to some. Climate change worsens health inequities, disproportionately harming the most vulnerable among us—children and pregnant women, low-income people, the aged and people with disabilities and chronic illnesses, some communities of color, indigenous people and tribal communities, immigrants, marginalized people of all races and ethnicities, and outdoor workers. Communities that have experienced systemic neglect and environmental racism have contributed little to climate pollution, but they are the most affected. These communities have less access to the political, economic, social, and environmental resources that enable them to cope with climate threats and face potentially unmanageable pressures as the impacts of climate change accumulate.

People's choices will determine how much climate impacts our children and grandchildren, and whether future generations will have access to the natural resources and environments that will enable them to be healthy. If we fail to take action now, options for limiting global warming and preventing catastrophes will no longer be available. U.S. climate policies and investments must right existing wrongs and address our moral responsibility to current and future generations.

Without action, climate change will be increasingly more severe, leading to more illness, injury, and death; mass migration and violent conflict; and worsening health problems. By mobilizing climate action for health and health action for climate, the United States can reduce pollution and build healthy communities that can stand up against climate risks.

We need to honor commitments to climate action and accelerate action to protect our health and the health of future generations. With the right policies and investments today, we can realize our vision of healthy people in healthy places on a healthy planet. The priority actions outlined below are urgent and essential steps to protecting and promoting the health and well-being of all people in the era of climate change.

Climate Action for Health

Making health essential to climate policymaking at all levels and across all sectors allows greater support for climate action and an opportunity to advance climate solutions and to achieve ambitious health targets through win-win strategies that promote climate justice, health and health equity, resilience, and a sustainable economy. We urge government leaders to advance the following priorities.

Priority Actions

1. *Meet and strengthen U.S. commitments under the Paris Agreement. A large and rapid reduction in carbon emissions is essential for our health and the health of future generations. The United States must recommit to the Paris Agreement and to aggressive emissions reductions sufficient to limit a global temperature increases to 1.5°C above preindustrial levels and to continue to engage with international and national leaders, business, and civil society to encourage and support others to develop multilateral, binding commitments to do the same. The United States must ratify and implement the Kigali Amendment to reduce the use of hydrofluorocarbons.*

2. *Transition rapidly away from the use of coal, oil, and natural gas to clean, safe, and renewable energy and energy efficiency. With the technology available today, we can dramatically change U.S. energy use and systems to meet growing energy needs affordably, while reducing climate and air pollution. Key policies include:*
   - *Establish ambitious goals and timelines for renewable energy, energy efficiency, and energy conservation.*
   - *Support financing for the technologies and infrastructure needed to transition to zero carbon emissions, including development, adoption, and scale-up of renewable energy sources and investments in energy efficiency. Put a price on carbon that reflects its true social costs and phase out investments in and subsidies for fossil fuels for energy extraction and generation.*
   - *Ensure that climate policies support sustainable energy for all by promoting distributed renewable energy and zero emission transportation technologies, with a priority on disadvantaged communities.*
   - *Support a rapid reduction of petroleum and natural gas use in transportation through steady investment and regulations to increase fuel efficiency and transition to zero emission vehicle technologies as quickly as possible across the transportation sector.*
   - *Establish ambitious goals for building efficiency and move toward a zero carbon future by reducing carbon impacts from new and existing buildings. Transition away from wood burning, oil, and natural gas use for home heating and cooking.*
   - *Reduce conventional air pollutants along with reductions in carbon and short-lived climate pollutants to maximize health benefits in communities impacted by pollution.*
   - *Assess and address the health impacts of fossil fuels (coal, oil, and gas) extraction, production, transport, and infrastructure on urban and rural communities, for example, through "setbacks" for sensitive populations and stronger protections against fossil fuel industry impacts on clean air and water.*
   - *Develop a plan and timeline for reduction of fossil fuel extraction in the United States.*
   - *Support research on strategies to draw down climate pollution from the atmosphere and store it in the ground, and on the potential health and equity impacts of these strategies.*

3. *Emphasize active transportation in the transition to zero-carbon transportation systems. Shifting from driving to active modes of travel—walking, bicycling, and public transit—can substantially reduce rates of noncommunicable diseases (e.g., obesity, cardiovascular disease, diabetes, osteoporosis), and injuries. Key policies include:*
   - *Make transportation carbon reductions central to the mission of transportation agencies, and align transportation expenditures with the goals of reducing*

climate pollution and vehicle miles traveled and supporting healthier communities and travel choices for all.

- *Significantly increase the percentage of transportation investments for infrastructure and programs to promote safe walking and cycling, and for affordable, accessible, and convenient public transit infrastructure, maintenance, and operations including in rural communities.*
- *Invest in affordable housing to avoid displacement and very long-distance commuting based on families' ability to afford housing near jobs.*

4. *Promote healthy, sustainable and resilient farms and food systems, forests, and natural lands. By changing what we eat, and how we grow, harvest and transport our food, we can protect our health, reduce obesity, diabetes, and heart disease, and significantly reduce our carbon footprint. Properly managed and protected forests, farms, rangelands, and wetlands can serve as resilient carbon sinks and protect the communities that depend on them from climate impacts. Practices that reduce food waste, conserve and regenerate our soil, conserve and protect our water, sustain our fisheries, conserve productive agricultural land from urban sprawl, and protect those who grow our food are essential to safeguard our food supply and our safety in the face of climate impacts. Building resilient, ecologically sustainable, local food systems can support the livelihoods of agricultural communities and the people that grow and produce our food, expand access to healthy food, improve air and water quality and biodiversity, and reduce carbon emissions. Key policies include:*
   - *Invest in programs and encourage practices that protect, manage, conserve, and expand natural and working lands to increase carbon sequestration and reduce catastrophic wildfires, floods, and mudslides.*
   - *Expand tree canopy, parks, green spaces, and green infrastructure to sequester carbon, increase cooling in urban areas and reduce the impacts of flooding.*
   - *Use agricultural funding and programs to prioritize and enable a rapid shift to diversified and sustainable agro-ecological and regenerative practices that reduce reliance on chemical- and energy-intensive industrial monoculture and animal-based agriculture and environmentally damaging agricultural and fisheries practices. Support urban and periurban agriculture.*
   - *Integrate urban and agricultural land use planning to maximize transit-oriented infill development while conserving productive agricultural lands on urban edges.*
   - *Establish incentives and supports for reduction of food waste.*
   - *Incentivize livestock manure management practices that reduce potent methane emissions and produce valuable compost for soil fertility.*
   - *Encourage America's children to enjoy healthy, plant-based diets and reduce consumption of red and processed meat by implementing a strategy to provide meat-free options in all school meals.*

5. *Ensure that everyone in the United States has access to safe and affordable drinking water and a sustainable water supply. There is nothing more fundamental to human existence than water. Key policies include:*
   - *Enhance regulations to prevent water contamination from agricultural, mining, industrial, and energy production sources.*
   - *Invest in programs for water conservation and efficiency, water resources management, infrastructure maintenance, protection from flooding and salt-water inundation, and in research on sustainable and ecologically safe alternative water resources such as desalination and reuse.*

6. *Invest in policies that support a just transition for workers and communities adversely impacted by climate change and the transition to a low-carbon economy. A sustainable and equitable low-carbon economy requires shared prosperity, including fair employment and economic opportunities for workers and communities that are affected by climate change and climate-related policies and programs. Investment in green jobs builds community economic well-being and improves health. Key policies include:*
   - *Assess and alleviate impacts on workers and communities affected by job or economic losses related to climate change and climate policy, using inclusive engagement with stakeholders.*
   - *Advance a just transition through greater investments in workforce training and development, local hiring programs, and community-driven infrastructure.*

Health Action for Climate

Proactive support is required to expand health sector efforts to reduce greenhouse gas emissions in health facilities; build resilience through the integration of climate considerations in health systems, policies, programs, and investments; and effectively communicate the health threats of climate change together with the health benefits of climate action.

Priority Actions

7. *Engage the health sector voice in the call for climate action. Proactive health sector leadership in climate communications can significantly increase public support for transformative climate action. A key policy includes:*
   - *Implement local and national campaigns, using lessons from public health campaigns such as tobacco control, to inform about the health impacts of climate change and the health benefits of climate action.*
8. *Incorporate climate solutions into all healthcare and public health systems. Public health agencies must address climate change as a health emergency to protect and promote the health of communities. Hospitals and healthcare systems must implement climate-smart health care, build facility resilience, and leverage their economic power to decarbonize the supply chain and promote equitable local economic development. Key policies include:*
   - *Proactively support integration of climate change into all relevant federal, state, and local public health programs.*
   - *Establish a public-private task force to assess the current state of the nation's healthcare system's resilience to extreme weather and recommend strategies and investments to improve it.*
   - *Support policies to advance implementation of climate-smart energy, water, transportation, food, anesthetic gas, and waste management practices in U.S. healthcare facilities, including clinics and provider offices.*
9. *Develop low-carbon healthcare delivery models, utilizing community-based care sites, telemedicine and mobile technologies.*
   - *Support redesign of all health professional curricula to better prepare the health workforce to lead in climate change mitigation and adaptation.*
10. *Build resilient communities in the face of climate change. Climate change is a global phenomenon, but it is people and communities at the local level that experience its consequences. Climate and health action will be most effective when those most*

impacted have the voice, power, and capacity to be full partners in building a healthy, equitable, and climate resilient future. Key policies include:

- *Deeply engage communities most impacted by climate change and poor health outcomes in planning, policy development and budgeting, offering meaningful roles and power in decision-making processes, and respecting history, traditional ecological knowledge, and community-directed solutions.*
- *Support adequate planning and funding to protect all communities from the adverse health impacts of climate change, including robust heat island mitigation; expansion of tree canopy, green space, and green infrastructure; cool roofs and cool pavements; rainwater and gray water capture; strategies to reduce the occurrence and impacts of catastrophic wildfires and floods; community preparedness and resilience training; and increased availability of climate-adapted housing.*
- *Integrate and provide guidance on assessment of the health and health equity benefits (or risks) of proposed climate solutions and investments.*

Financing Climate Action for Health and Health Action for Climate

Achieving goals for climate, health, and equity will require that climate investments consider health impacts and benefits, and that investments in health take climate change considerations into account. Investing in the health of people and our communities saves money over time and makes the nation stronger. Current investments fall far short.

Priority Action

11. *Invest in climate and health.*
    - *Allocate resources to enable the health sector to effectively protect health in the face of climate change, starting with support for local and state health departments and a resilient hospital infrastructure.*
    - *Fund and implement national, state and local climate-health risk assessments, expanded disease surveillance systems, early warning systems, and research on climate and health that enable an effective health response to climate threats. Make all data publicly available.*

Summary

Together, these 10 policy recommendations provide a roadmap to develop coordinated strategies for simultaneously tackling climate change, health, and equity. Climate change is a health emergency. We call on local, state, and national leaders to act now to stop climate pollution, promote resilient communities, and support healthy people in healthy places on a healthy planet.

Policy Efforts to Date

Although a comprehensive discussion of these policies is beyond the scope of this chapter or this primer, several policy approaches currently in use around the United States are discussed here.

Federal strategies for moving the country in a new direction include international treaties such as the Paris Climate Treaty, the power to tax, monetary incentives (related to the power to tax), federal regulations through federal agencies (Departments of Energy and Agriculture, Environmental Protection Agency [EPA]), promotion of ideas through education, hearings, and programs administered by the many agencies of government. State governments use other strategies. Some of the strategies in use are presented here.

Federal and state policies have addressed some of the suggested targets. Policies that address climate change actually began with efforts to address air pollution, but the policies that have been explicitly focused on climate change to date have mostly addressed energy policy and transportation. Examples are regulating vehicle emissions, renewable energy standards and incentives for increasing the use of clean, renewable power for generating electricity, public transportation projects, and improved energy efficiency. Local jurisdictions use other approaches since they have the responsibility for local zoning and building codes that have significant implications for use of electric power, and energy efficiency. Since these requirements can change the quality of outdoor and indoor air, they can also affect health.

Private-sector businesses and nonprofit institutions such as universities and hospitals are reducing energy waste, improving energy efficiency, employing clean, renewable energy, purchasing from sustainable sources in their supply chain, and purchasing agricultural commodities that are sustainably sourced and local.

Regulation of Air Pollution

Governmental policies that address emissions of fossil fuels as air pollutants have been implemented across America for more than 50 years. At the advent of these policy developments, climate change was not on the radar screen of policymakers. The concern at that time was the health impact and environmental damage that accompanied harmful discharges into the air. These policies were initiated by an act of Congress but have continued on the basis of regulatory actions implemented by agencies and shaped by scientific analysis and medical research.

The Clean Air Act was passed in 1970. This legislation established the foundation for a regulatory approach to improving air quality. On this foundation, pollution limits and requirements for low-emission vehicles and fuel economy (using less fuel for the same energy generation) followed. Policies were enacted at the national and state levels. Because this policy addresses emissions of greenhouse gases, it has significant implications for climate change. It illustrates the close connection between the use of fossil fuels and health at the level of direct exposure to pollution in addition to the macro level in which carbon pollution has accumulated to the point where it adversely affects our atmosphere and is causing global warming and climate change.

The Regional Greenhouse Gas Initiative (RGGI). The RGGI is another regulatory approach established at the state level in the northeast in 2009 specifically for the purpose of addressing climate change. The number of states that voted to join the RGGI grew to 10 as of 2020, but nine states in the northeast came together in 2009 to reduce CO_2 emissions whose source was electric power production from the burning of fossil fuels. They created a carbon market called the Regional Greenhouse Gas Initiative (RGGI), which is a "**cap and trade**" system that places a regional limit or cap on the amount of CO_2 that power plants can emit. This action was like a treaty between the states, approved by their legislatures and governors. Any power plant that emits more than the cap pays a penalty. The "trade" means that businesses that are not emitting up to their cap may trade with others and sell the right to emit CO_2. The cap is lowered over time so that the total amount of CO_2 going into the atmosphere is less. In addition to CO_2, since other pollutants are emitted when fuel is burned for power, the RGGI reduces other pollutants too. The typical pollutants of concern are CO, ground-level ozone, nitrogen oxides, particulate matter, and SO_2, and lead. These are mentioned in the original Clean Air Act discussed below and referred to as "**criteria pollutants**" since they are regulated according to specific criteria and levels by the federal Environmental Protection Agency (EPA). Since burning coal releases more CO_2 than burning natural gas (methane), the cap and trade system applies financial pressure to reduce coal use and make all power generation cleaner and more efficient. Burning coal releases toxic metals like cadmium, mercury, and arsenic into the air. If the cap and trade system reduces use of coal burning over time, these pollutants will also decrease.

There is concern about the impact of the RGGI system on communities that have received a disproportionate share of the burden of air pollution. A system-wide reduction in emissions of pollutants may not reduce the dirty air burden on communities that have poor local air quality. Health professionals should judge policies by the potential to benefit health and to impact the social determinants of health as well as the potential to reduce greenhouse gas accumulation. Work will undoubtedly continue to address this environmental injustice because it has far reaching affects on the health of the persons who are exposed.

In late 2018, the RGGI states announced that they would establish another RGGI system that specifically addresses transportation. This would set up a market that applies to CO_2 that is released when fossil fuels are burned in transportation. They are hoping that such a system will accelerate the introduction of nonpolluting electric cars. Critics of this approach point out that cleaner cars are important but do not address some of the other problems associated with the car culture, namely traffic, congestion, and vehicular deaths. The sheer magnitude of our landscape given over to roads is another criticism of this approach. It is argued that with more pedestrian friendly communities, business and social relationships have more room to flourish; cooling greenery like trees and shrubs may be planted over more territory as well.

A January 2017 study by Manion and colleagues that focused on the environmental and public health impact of the RGGI found that it "created major benefits to public health and productivity, including avoiding hundreds of premature deaths and tens of thousands of lost work days."[29] Specifically, this report found that RGGI annually prevented the following:

- 300–830 early deaths among adults
- 35–390 nonfatal heart attacks
- 8,200–9,900 asthma flare-ups
- 200–230 asthma ER visits
- 180–220 hospital admissions
- 39,000–47,000 lost work days

The total health cost savings from the Regional Greenhouse Gas Initiative (2009–2017) is estimated at $5.7 billion. These are undoubtedly some of the achievements that public health and medical professionals want from a government policy intended to address climate change: fewer people going to hospitals, fewer people missing work, fewer people seeking care because they suffer from serious health conditions, and, of course, fewer people dying prematurely. However, the additional concern about addressing the needs of the people and communities already impacted by environmental justice is real and should be factored into policy development.

At the federal level, a cap and trade system almost became law at the federal level during the 111th Congress in 2009–2010. The system was introduced into both houses of the United States Congress in 2009 as the "American Clean Energy and Security Act," which had several provisions to address climate change. It passed the House and nearly passed the Senate with 59 Senators supporting it. However, 60 votes were needed in the Senate to prevent a filibuster. Ultimately, the majority could not convince the 60th Senator to support this plan. The Bill would have reduced greenhouse gas emissions by 83% by 2050. Other policy provisions addressed transitioning to a clean energy economy, providing for agriculture and forestry-related offsets in which additional plants would reduce the CO_2 in the air, and creating energy efficiency and renewable energy standards. There was also a requirement that retail electricity suppliers meet 20% of their demand through renewable electricity and electricity savings by 2020. Although the House of Representatives passed this package of policies, and a Senate majority supported it (59 votes), this work remains unrealized.

The Clean Air Act

This landmark federal legislation passed in 1970, built on prior legislation from twenty years previously that gave the federal government the right to monitor, inspect, and regulate pollution and support research on its effects. The 1970 act expanded the federal mandate, requiring comprehensive federal and state regulations for both stationary sources of pollution and mobile sources (vehicles) and significantly expanded federal enforcement. Specifically, the Clean Air Act of 1970:

- Established National Ambient Air Quality Standards
- Established requirements for State Implementation Plans to achieve them
- Established New Source Performance Standards for new and modified stationary sources
- Established the National Emission Standards for Hazardous Air Pollutants
- Increased enforcement authority
- Authorized control of motor vehicle emissions

The pollutants covered by the law are referred to as "criterion pollutants," specifically ozone, particulate matter, carbon monoxide, nitrogen oxides, sulfur dioxide, and lead (CO_2 was added later; see the section on the Endangerment Finding below). The EPA was also created in 1970 by President Nixon to administer these authorities and consolidate federal research, monitoring, standard-setting, and enforcement into one agency that ensures environmental protection.

Addressing air pollution is another way to approach climate and health policy. Since drilling and burning fossil fuels creates air pollution, which is harmful to health, addressing air pollution is also an effective way to address climate change. At this time, more is known about the effect of burning these fuels than drilling for them. Research has also started to emerge on the harmful effects of drilling for these fuels, especially for those who live close to these operations—and especially for pregnant women and children. Furthermore, evidence of the direct health benefits of reducing air pollution has grown substantially over the years.

Table 4-1 Health Benefits of the Clean Air Act

	Year 2010 (cases prevented)	Year 2020 (cases prevented)
Adult Mortality—particles	160,000	230,000
Infant Mortality—particles	230	280
Mortality—ozone	4,300	71,000
Chronic Bronchitis	54,000	75,000
Heart Disease—Acute Myocardial Infarction	130,000	200,000
Asthma Exacerbation	1,700,000	2,400,000
Emergency Room Visits	86,000	120,000
School Loss Days	3,200,000	5,400,000
Lost Work Days	13,000,000	17,000,000

The benefits reaped from the Clean Air Act, from 1990 to 2010, were summarized by the EPA in 2011.[30] **Table 4-1** shows the health benefits of the Clean Air Act due to reductions in fine particle and ozone levels. The return on investment is estimated as 30 dollars saved for every dollar spent. That is the average estimate; the lowest estimate is 3 to 1.[31]

The results of the research that provides the basis for regulatory policy continues. Recent evidence shows other impacts of air pollution on the growth and development of children's brains and lungs. While much is known about how polluted air affects the lungs and heart, we are just beginning to learn about how it can also harm the brain. Over the last decade, many studies have linked outdoor air pollution exposure to harmful impacts on the brain.[32]

Clean Car Standards

The 1970 Clean Air Act also gave the EPA the authority to regulate mobile sources or motor vehicle emissions. Figure 4-1 at the beginning of this chapter shows that transportation is one of the two largest contributors of greenhouse gases. Reduction of emissions from mobile sources has helped to reduce particulate and ozone levels significantly over the years. The U.S. Office of Management and Budget, which is within the Office of the President and prepares the federal budget, estimates that every $1 spent on auto emission reductions has saved $9 in public health, environment, productivity, and consumer savings.[33] The EPA has coordinated with the U.S. Department of Transportation so that standards are also set for fuel efficiency, assuring a smaller amount of burning and greenhouse gas products from vehicles.

Both health and economic benefits result from clean car standards. Longitudinal studies have been conducted on cohorts of children in southern California who grew up under increasing limits on air pollution (mainly from vehicles). These studies show that the lung function of children (ages 8 to 15) is better if they grow up in less polluted air and it is worse if they grow up in more polluted air. The principal pollutants associated with reduced lung function identified in these studies were tiny particles called "particulates." Children with less exposure to particulates have better functioning lungs. Ozone is another powerful irritant of the mucous membranes in the lungs and eyes. Much research supports the connection between the health burden and air pollutants, as well as the association between infant health and reduced traffic.[34,35]

Economic analyses confirm these findings by documenting increased health costs for children who live in areas with more fine particulate matter.[36] According to a study published in the health policy journal *Health Affairs* in 2011, in an issue devoted to health and the environment, if the United States reduced levels of fine particulate matter by 7% below the current national standards, the country could save another $15 million per year from reduced hospitalization in urban areas for bronchiolitis. These findings reinforce the need (and the health and economic benefits) for ongoing efforts to reduce levels of air pollutants."[37]

Clean Power Plan

At times, progress is stymied by other policies, use of the courts, or both. In 2014, a new federal regulatory structure called the Clean Power Plan was announced by the EPA that required states to reduce the CO_2 output from their energy-generating power plants. The target was a 32% reduction by 2030. Incentives for creating clean, renewable energy sources were part of the plan. Coal plants were also required to pursue carbon sequestration to prevent the harmful emissions that come from burning coal. Coal burning leads to greater CO_2 output than burning other forms of fuel. After a short time period, the Clean Power Plan was finalized as law in June of 2015.

A few months later, 27 states and some businesses went to court and asked that this regulation be stopped because it was financially harmful. Due to earlier Supreme Court decisions in 2009 and 2014, and the "Endangerment Finding," the regulation could not be challenged on the claim that the EPA did not have the right to regulate it. The case went to the Supreme Court, which ruled 5–4 that the Clean Power Plan should be placed on hold while it worked its way through the lower courts.

The year after he was elected, President Trump announced that the EPA would abandon and replace the Clean Power Plan. The EPA announced a new plan, the Affordable Clean Energy Plan, in 2018, which allows states to set their own pace with no specific targets. If this passes through the regulatory hearings and review, the new approach will be ready for implementation. There will likely be a legal challenge to this as well.

The legal basis for the Clean Power Plan came from a prior court case and an EPA finding that was affirmed by the Supreme Court. In 1999, Massachusetts and 11 other states sued the EPA (*Massachusetts v. EPA*) because they were concerned about the impact of environmental greenhouse gases, including CO_2 and others, on climate change and the nation's health. In 2007, the decision was finally announced when the Supreme Court held that greenhouse gases are pollutants under the Clean Air Act. The Court focused on the provisions of the Clean Air Act that require regulation when the EPA finds that emissions of a pollutant endanger public health or welfare. The EPA was required to establish whether greenhouse gases and CO_2 were a danger to public health or welfare.

Based on a careful review of the scientific record, this finding was issued at the end of 2009 when the EPA determined that greenhouse gas emissions endanger the public health and welfare of current and future generations. This is referred to as the "endangerment finding." There was extensive scientific evidence referencing 100 published scientific studies, peer-reviewed syntheses from the US Global Change Research Program, the National Research Council, National Academy of Sciences, and the United National Intergovernmental Panel on Climate Change. There were also solid testimonies from private citizens. The EPA work on this occupies 11 volumes and over 500 pages. There was a lot of opposition to this conclusion.

The Endangerment Finding

The Endangerment Finding that human activity is generating the greenhouse gas pollution that is causing climate change—was upheld by the D.C. Circuit court and then by the Supreme Court (2014). Briefs from 16 states supported the case. In this decision, the Court left undisturbed key Clean Air Act provisions authorizing the EPA to issue "performance standards" that limited carbon pollution from sources such as power plants, refineries, and cement kilns. It also preserved the EPA authority to limit carbon pollution from cars and trucks.

Renewable Energy Portfolio Standards (RPS)

Using "**renewable energy portfolio**" or renewable energy standards is a legislative strategy that is only at the state level so far. It signifies a legislative decision to derive a specific percentage of the energy produced in a state jurisdiction from clean, renewable sources, such as solar power, wind power, hydroelectric power, geothermal energy, or some other source. Since the early 2000s, 29 states, Washington, D.C., and three territories have adopted RPS, and eight other states and one territory set renewable energy goals.[38] Iowa was the first state to have a RPS and Hawaii has the largest requirement—100% renewable energy by 2045. Twenty-nine states and the District of Columbia have such standards.[39] These standards have played a role in driving the expansion of clean, renewable energy in the United States. In 2017, 17% of electricity generated in the United States was from clean, renewable sources.[40] Studies that examine the health impact of the Renewable Portfolio Standards are unavailable. This evidence will evolve over time.

State Moratoria on Drilling for Natural Gas (Methane)

Drilling for natural gas or methane is a private-sector activity typically regulated by states (unless it occurs on federal land) that raises air pollution and climate policy concerns. As mentioned earlier, methane is many times more potent as a greenhouse gas (GHG) than CO_2. Burning methane produces less CO_2 than coal, and some feel it is an essential part of the energy mix. They are encouraged that using natural gas has increased as much as it has in the United States. However, methane escapes into the atmosphere at drilling and transfer points and adds substantially to the warming of the atmosphere. Because a great deal of methane is frozen into the melting permafrost lands of the arctic regions of the Earth, there will be an even more dangerous release of methane over time.

There are other issues associated with drilling for natural gas. Some of the direct impacts of exposure for people who live at the "fenceline" was mentioned previously. Furthermore, since much of the natural gas is drilled using fracking techniques where large amounts of water and dangerous chemicals are forced deep into the ground to push the natural gas upward, dumping the contaminated water and chemicals that are recovered at the end of the process is a concern. It can end up in the waterways that provide drinking water or connect to other waterways that are drinking water sources. In some places, the natural gas has migrated underground into well water that provides drinking water for local residents. Some residents have noticed a strange smell from their tap water. They found that the water had so much gas in it that they could light it on fire! Because of such concerns, New York, Vermont, and Maryland banned fracking.

Carbon Taxes or Carbon Fee and Dividends

Another policy approach is the carbon tax, which uses the Internal Revenue Service's infrastructure to tax every metric ton of CO_2 that is emitted. At this time, a carbon tax is in effect in 10 developed countries and several subnational provincial jurisdictions. While some people in the United States oppose this approach because it is a tax, other issues must also be resolved for this to go forward. The revenue that is raised by this type of system can be spent in different ways, including a return to all taxpayers, protecting communities that are vulnerable to climate change impacts, or a return to the general treasury. If a carbon tax is established, there will be choices made about use of the revenue.

The Role for Individual Behavior Change

In addition to the national and state-level policies and regulations mentioned above, many people are personally trying to help reduce the growth of the greenhouse gas (GHG) concentration in the Earth's atmosphere. Most individual behavior adjustments that contribute to lowering carbon dioxide output are also beneficial for personal health.

- Choosing clean, renewable power if it is available from an electricity supplier where you live decreases the pollution that is being added to the air that everyone breathes. Saving energy by using less or making your home more energy efficient also reduces energy waste and the pollution that is added to the air when fuels are burned.
- Choosing more active transportation such as public transportation or alternatives such as walking or biking helps people get more exercise and benefits health. Selecting an electric car is another option.
- Eating more fruits vegetables and less meat reduces the agricultural contribution to GHG because meat, especially beef, accounts for a large percentage of it, and fruits and vegetables are part of a healthier diet.
- Eating local, organically grown fruits and vegetables helps because local food travels a shorter distance, resulting in fewer transportation-related emissions. The organic approach also puts the carbon that is in the plants back into the

ground where it is sequestered, and it requires less fertilizer, which requires the potent greenhouse gas methane to be produced and adds another GHG (NO_2) to the atmosphere.

Clearly, there is much that can be done on the personal level and the policy level to address the overload of greenhouse gases that is affecting our atmosphere and our climate. Everyone can contribute to addressing this serious problem that Americans and other people around the world are facing.

Key Reports
Draw Down

If the problem is viewed as the increased concentration of CO_2 already in the air, solutions might focus on how to "draw down" that CO_2. A book called "**Draw Down**" approaches the problem this way and rank orders many solutions that could help draw the CO_2 out of the atmosphere by putting it somewhere else or preventing its emission (as well as the emission of the other greenhouse gases). An extensive analysis shows that some solutions could have a significant impact.[41]

Intergovernmental Panel on Climate Change (IPCC)

This is the United Nations body for assessing the science related to climate change. The **IPCC** provides regular assessments of the scientific basis for climate change, its impacts and future risks, and the options for adaptation (protection) and mitigation (prevention through reducing greenhouse gas emissions). It was founded by the World Meteorological Organization and the United Nations in 1988 and now represents 195 countries. Through its evaluations, the IPCC identifies the strength of scientific agreement in different areas and indicates where further research is needed. The IPCC does not conduct its own research. The IPCC is divided into three working groups and a task force. Working Group I deals with The Physical Science Basis of Climate Change; Working Group II with Climate Change Impacts, Adaptation, and Vulnerability; and Working Group III with Mitigation of Climate Change. The main objective of the Task Force on National Greenhouse Gas Inventories is to develop and refine a methodology for the calculation and reporting of national greenhouse gas emissions and removals. This task is basic to any international agreement to address the problem.[42]

The Lancet Count Down: Tracking Progress on Health and Climate Change

The Count Down is an international research collaboration, dedicated to tracking the world's response to climate change, and the health benefits that emerge from this transition. Reporting annually *The Lancet* follows a series of indicators, demonstrating the health impacts to date, and that this transition is possible, that it has already begun, and that more work is needed. There is a U.S. Policy Brief, which is developed and shared annually by U.S. policy experts that highlight specific indicators for the United States.[43]

National Climate Assessment (NCA)

The Global Change Research Act of 1990 was passed by Congress and requires a state-of-the-science report every four years on climate change. These reports are referred to as National Climate Assessments, and they synthesize climate impacts and trends across U.S. regions and sectors. The law gave this responsibility to the federal Global Change Research Program, an interagency working group representing 13 federal agencies. This program organizes prominent experts and manages the report, assuring review by a broad cross section of federal agencies and a panel of the National Academy of Sciences. The most recent National Climate Assessment, NCA4, the fourth in the series, was released in two parts: Volume I released in November 2017 and Volume 2 released in November of 2018. NCA4 Volume I (*Climate Science Special Report*) covers observed and projected changes in the physical climate system. Volume II (*Impacts, Risks, and Adaptation in the United States*) focuses on climate-related risks to humans and systems that support our well-being and economy.[44]

Special IPCC Report on Global Warming of 1.5 Degrees

An IPCC special report was released in 2018 on the impact of global warming of 1.5°C above pre-industrial levels. This report was requested as part of the decision of the 21st Conference of Parties of the United Nations Framework Convention on Climate Change to adopt the Paris Agreement in 2015. The agreement was to hold global warming to 2.5 degrees and make the best efforts to hold it to 1.5 degrees. The special report describes the impact of global warming of 1.5°C above preindustrial levels and related global greenhouse gas emission pathways in the context of strengthening the global response to the threat of climate change, sustainable development, and efforts to eliminate poverty.[45]

United Nations Framework Convention on Climate Change

This is an international environmental treaty first adopted in 1992 at the Earth Summit in Rio de Janeiro, Brazil, becoming effective in 1994 after a sufficient number of countries had signed it. The objective is to "stabilize greenhouse gas concentrations in the atmosphere at a level that would prevent dangerous anthropogenic interference with the climate system."[46]

Recap

Earlier in this Primer, we learned that climate change is harming the health of people today throughout the United States—and indeed the world; that it is, harming some more than others (children, pregnant women, people with chronic conditions, poor people, and communities facing environmental injustice); that it will get worse if we don't address the problem; and that addressing the problem can be done by transitioning to clean, renewable sources of energy and embracing energy efficiency in the generation of energy, in transportation, and building design. Improvements in agriculture, climate smart health care which produces a smaller environmental footprint, and a focus on transitions for affected communities are also needed. The sooner we make these changes, the sooner we will have cleaner air and water and better population health. Understanding how and why climate change is happening, the current array of potent greenhouse gases and the sources of those gases, and the policy roadmap for addressing climate change that has been endorsed by numerous health professional groups, schools, and health systems gives all health professionals the background, understanding, and policy prescriptions to contribute to making climate change a national policy priority. This will benefit everyone's health right now, today, and for generations to come.

WRAP-UP

Discussion Questions

1. Should health be the principal basis on which to evaluate policy solutions to climate change? Why or why not? What other yardsticks might people use to evaluate alternative climate policies?
2. Coal received early and strenuous opposition as a source of energy for heating and cooling in the Unites States. Discuss the factors that make coal especially problematic as a fossil fuel source of energy.
3. The level of concern about living in proximity to fossil fuel drilling sites has grown with time. How do you explain the increasing alarm about exposure to these sites?
4. Which of the policies in the Policy Action Agenda are likely to have the greatest impact on climate change? Explain your reasoning.

References

1. Hoffman JS, Shandas V, Pendleton N. The effects of historical housing policies on resident exposure to intra-urban heat: a study of 108 US Urban Areas. *Climate.* 2020;8(1):12.
2. Gauderman WJ, et al. Association of Improved Air Quality with Lung Development in Children, nejm.org, March 5, 2015.
3. Conticino E, Frediani B, Caro D. Can Atmospheric Pollution be Considered a Co-factor in Extremely High SARS-COV 2 lethality in Northern Italy? Environmental Pollution. In press.
4. APHA.org. Energy justice and climate change: key concepts for public health. 2020. *Fact sheet on energy justice and key concepts for public health professionals from the American Public Health Association.*
5. Fleishman L, Franklin M. Fumes Across the Fence-Line: The Health Impacts of Air Pollution from Oil and Gas Facilities on African American Communities. NAACP Clean Air Task Force and National Medical Association. November 2017. https://www.naacp.org/wp-content/uploads /2017/11/Fumes-Across-the-Fence-Line_NAACP-and-CATF-Study.pdf. Accessed July 26, 2020.
6. U.S. Energy Information Agency. What is U.S. electricity generation by energy source? 2020. https://www.eia.gov/tools/faqs/faq.php?id=427&t=3b. Accessed December 24, 2019.
7. Dalesio EP. Duke energy sued for 2014 coal ash spill environmental harm. *AP News.* July 18, 2019. https://www.apnews.com/d669e8e278ca47e4997088d0f798e583. Accessed October 13, 2019.
8. Currie J, Greenstone M, Meckel K. Hydraulic fracturing and infant health: new evidence from Pennsylvania. *Science Advances.* 2017;3(12):e1603021.

9. Landrigan PJ, Fuller R, Acosta NJ, et al. The *Lancet* Commission on pollution and health. *Lancet.* 2017;391(10119):462-512.

10. Bennett D, Bellinger DC, Birnbaum LS, et al. Project TENDR: targeting environmental neuro-developmental risks. The TENDR Consensus Statement. *Environmental Health Perspectives.* 2016;124(7).

11. Currie J, Greenstone M, Meckel K. Hydraulic fracturing and infant health: new evidence from Pennsylvania. *Science Advances.* 2017;3(12):e1603021.

12. Casey JA, Karasek D, Ogburn EL, et al. Retirements of coal and oil power plants in California: association with reduced preterm birth among populations nearby. *Am J Epidemiol.* 2018; 187(8):1586-1594.

13. Bekkar B, Pacheco S, Basu R, Denicola N. Association of air pollution and heat exposure with preterm birth, low birth weight, and stillbirth in the US. A systematic review. *JAMA Network.* 2020:1-13.

14. Hayes S, Kubes C. Saving Energy saving lives: the health impacts of avoiding power plant pollution with energy efficiency. 2018.

15. U.S. Green Building Council. Public Policy Library. (nd) https://public-policies.usgbc.org/. Accessed June 1, 2020.

16. https://www.usgbc.org/resources

17. Eckelman MJ, Sherman J. Environmental impacts of the U.S. health care system and effects on public health. *PLOS One.* 2016;11(6):e0157014.

18. Sherman JD, MacNeill A, Thiel C. Reducing pollution from the health care industry. *JAMA.* 2019;322(11):1043-1044.

19. Practice Greenhouse. Sustainability solutions for health care. 2020. https://practicegreenhealth.org/

20. Cobiac LJ, Scarborough P. Modelling the health co-benefits of sustainable diets in the UK, France, Finland, Italy and Sweden. *Eur J Clin Nutr.* 2019;73(4):624-633.

21. Healthy Diet Sustainably Produced. World Health Organization. 2018.

22. Healthy diets from sustainable food systems: summary report of the EAT-Lancet Commission. *EAT.* 2019.

23. Future 50 Foods Report. World Wildlife Fund, Feb. 2019.

24. Associated Press. Climate changes is a 'health emergency,' 74 medical and public health group warn. 2019. https://www.nbcnews.com/health/health-news/climate-change-health-emergency-74-medical-public-health-groups-warn-n1020961. Accessed June 18, 2020.

25. Climate Health Action.org. U.S. Call to action on climate, health, and equity: a policy action agenda. 2019. 1-6. https://climatehealthaction.org/media/cta_docs/US_Call_to_Action.pdf. The document was developed by Dr. Linda Rudolph in consultation with several health organizations.

26. Climate Action Summit. The Climate Group. September 2018. https://www.theclimategroup.org/event/global-climate-action-summit. Accessed on July 26, 2020.

27. World Health Organization. COP24 Special report: Health & Climate Change Report: 3. 2018.

28. Carmona R, Satcher D. Climate Change is Affecting Health Now: Our Leaders Must Take Action. STAT January 16, 2019. https://www.statnews.com/2019/01/16/climate-change-affecting-health-leaders-must-take-action/. Accessed July 26, 2020.

29. Manion M, Zarakas C, Wnuck S, et al. Analysis of the public health impacts of the regional greenhouse gas initiative, 2009–2014. 2017. http://abtassociates.com/AbtAssociates/files/7e/7e38e795- aba2-4756-ab72-ba7ae7f53f16.pdf. Accessed February 20, 2017.

30. United States Environmental Protection Agency. Clean Air Act Overview: progress cleaning the air and improving people's health. (n.d.). https://www.epa.gov/clean-air-act-overview/progress-cleaning-air-and-improving-peoples-health. Accessed December 27, 2018.

31. United States Environmental Protection Agency. The benefits and costs of the Clean Air Act from 1990 to 2020. 2011. https://www.epa.gov/sites/production/files/2015-07/documents/fullreport_rev_a.pdf. See the Abstract. Accessed December 27, 2018.

32. The link between fossil fuels and neurological harms. https://www.hbbf.org/sites/default/files/documents/2018-06/LinkFossilFuelsNeurologicalHarm_04-06-18_v2.pdf. Accessed June 19, 2020.

33. United States Environmental Protection Agency. History of reducing air pollution from transportation in the United States. 2020. https://www.epa.gov/transportation-air-pollution-and-climate-change/accomplishments-and-success-air-pollution-transportation. Accessed December 27, 2018.

34. Kheirbek I, Haney J, Douglas S, et al. The contribution of motor vehicle emissions to ambient fine particulate matter public health impacts in New York City: a health burden assessment. *Environ Health*. 2016;15(89).

35. Currie J, Walker R. Traffic congestion and infant health: evidence from E-ZPass. *American Economic Journal: Applied Economics*. 2011;3(1):65-90.

36. Roy A, Sheffield P, Wong K, et al. The effects of outdoor air pollutants on the costs of pediatric asthma hospitalizations in the United States, 1999–2007. *Med Care*. 2011;49(9):810-817.

37. Sheffield P, Roy A, Wong K, et al. Fine particulate matter pollution linked to respiratory illness in infants and increased hospital costs. *Health Affairs*. 2011;30(5):871-878.

38. National Conference of State Legislatures. State renewable portfolio standards and goals. 2020. http://www.ncsl.org/research/energy/renewable-portfolio-standards.aspx. Accessed December 26, 2018.

39. National Conference of State Legislatures. State renewable portfolio standards and goals. 2020. http://www.ncsl.org/research/energy/renewable-portfolio-standards.aspx. Accessed October 13, 2019.

40. Independent Statistics & Analysis. U.S. Energy Information Administration. What is U.S. electricity generation by energy source? (n.d.). https://www.eia.gov/tools/faqs/faq.php?id=427&t=3. Accessed December 26, 2018.

41. Hawken, Paul. Draw Down. Penguin Books 2017.

42. Intergovernmental Panel on Climate Change (IPCC). The intergovernmental panel on climate change. (n.d.). https://www.ipcc.ch/. Accessed January 9, 2019.

43. Lancet Countdown: Tracking Progress on Health and Climate Change: 2019 Report. 2019. http://www.lancetcountdown.org/. Accessed January 9, 2019.

44. National Climate Assessment. Climate change impacts in the United States. 2014. https://nca2014.globalchange.gov/. Accessed January 9, 2019.

45. Intergovernmental Panel on Climate Change (IPCC). Special report: global warming of 1.5°C. (n.d.). https://www.ipcc.ch/sr15/. Accessed January 9, 2019.

46. United Nations Climate Change. June Momentum for Climate Change. *The United Nations Framework Convention on Climate Change*. 2020. https://unfccc.int/. Accessed January 9, 2019.

Glossary

Adaptation The changes within living things and/or entire species that make it possible to live or survive more easily in the environment. Adaptation amidst climate change can help humans and other species blunt the disruptive impact of environmental changes.

Anthropogenic Factors originating from human activity, which may affect climate change.

Attention-Deficit Hyperactivity Disorder A developmental disorder marked by inattention and/or hyperactivity, which may have a causal association with exposure to air pollution, especially fine particulate matter.

Autism in Children A developmental disorder marked by difficulty socializing, communicating, and restrictive/repetitive patterns of thought and behavior, which may have a causal association with exposure to air pollution, especially fine particulate matter.

Black Lung A debilitating chronic lung disease that results from inhaling coal dust.

Cap and Trade A system that places a regional limit on the amount of CO_2 that power plants can emit. If an industry emits more than their limit, it pays a penalty. If a business emits less than their limit, they may sell their right to emit CO_2 to others.

Carbon Dioxide Concentration The amount of CO_2 by volume in the air. The Mauna Loa Mountain Observatory in Hawaii is the standard location for the measurement of CO_2 in the Earth's atmosphere.

Co-benefits The added benefits that may be achieved if climate change is controlled and protected. Generally used to refer to benefits to health.

Criteria Pollutants Pollutants that are indicators of air quality and regulated according to specific criteria by the federal Environmental Protection Agency (EPA) because of their impact on health and/or the environment.

Decreased Cognitive Function When there is a decreased ability to remember, learn, concentrate, or make decisions, which may have a causal association with exposure to air pollution, especially fine particulate matter.

Draw Down When carbon dioxide is taken out of the atmosphere and sequestered into the soil, trees, or other sinks.

Equity Equity exists when every person has what they need to live a happy and healthy life. It is different from equality in recognizing that each person or population has different needs, and some have been discriminated against and therefore kept from accessing basic goods, services, and societal benefits. Equity can mean providing more of something to one person or group so they have the same opportunities as someone else who doesn't need that support. An example would be building a ramp up to a door so that those in wheelchairs can access a building like those who are able to walk up the stairs.

Fenceline Community A population that lives close to fossil fuel extraction, refining, or storage process or sites, and which is thus exposed directly to harmful emissions. These populations tend to be disproportionately Black and Latino, and are more likely to live in constrained circumstances or poverty.

Fracking When liquids are injected at high pressure into subterranean rock to open fissures that allow the extraction of oil and/or gas.

Health Frame A way to analyze, discuss, and communicate a problem such as global warming by focusing on issues related to health.

Human Drivers Human caused forcing of atmospheric warming that include the heat absorbing and heat reradiating gases that humans have added to the atmosphere as products of human activity.

Insect Vectors Insect populations that carry infectious diseases.

IPCC The Intergovernmental Panel on Climate Change is a body of the United Nations that assesses the science related to climate change. It provides regular assessments of the scientific basis for climate change, its impacts and future risks, and the options for adaptation (protection) and mitigation (prevention through reducing greenhouse gas emissions).

LEED Certification An official and independent verification of Leadership in Energy and Environmental Design (LEED) provided by the U.S. Green Building Council to buildings or neighborhoods that are designed, built, operated, and maintained according to specific criteria and standards that are known to improve clean air and water, and the conservation of resources.

Mitigation Efforts to reduce the underlying causes of climate change (e.g. human drivers) and thus prevent continued damage to the life sustaining systems that support human life.

Natural Drivers Natural causes of atmospheric warming that include natural variations in the Sun's energy, changes in the path of the Earth revolving around the sun that brings the Earth closer to the Sun at certain times based on its elliptical orbit, and natural events on Earth, such as volcanic eruptions that produce aerosols that block the Sun's rays.

Neurodegenerative The degeneration of the nervous system, which may be caused by air pollution.

Neurodevelopmental The development and functioning of the nervous system, which can be damaged by air pollution.

Ocean Acidification The increasing amount of the acid H^+ and the molecule $^-HCO_3$ in the ocean.

Paris Treaty An agreement in 2015 by the parties of the United Nations Framework Convention on Climate Change to take actions that will combat climate change. In 2017 President Trump announced that the United States would abandon the agreement.

Redlining A term that references historically-discriminatory policies and practices that denied home loans, mortgage insurance, and investment (e.g. the building of a supermarket) in primarily Black, non-White and poor communities. Redlining practices were outlawed via the 1968 Fair Housing Act and 1977 Community

Reinvestement Act, but still have a deleterious impact on living circumstances, well being, and equity today.

Renewable Energy Portfolio A legislative strategy that signifies a decision to derive a specific percentage of the energy produced in a state jurisdiction from clean, renewable sources, such as solar power, wind power, hydroelectric power, or geothermal energy. This is also referred to as a "Renewable Energy Standard."

Rising Seas The increased volume of the oceans caused by such factors as the Earth's changing temperature and the melting ice sheets at the poles.

Segmentation A way of developing and testing different evidence-based strategies for communicating about something like global warming by dividing potential audiences into specific groups.

Self-efficacy A person's belief that they can or have the ability to do something, such as a behavior or action.

Sequester To capture, as when organic agriculture plows last year's plants back into the earth and carbon in those plants is captured in the ground.

Silicosis A disabling chronic lung condition that may result from inhaling dust that contains silica, a principal component of sandstone and other rocks.

World Health Organization (WHO) An agency of the United Nations whose stated main objective is "the attainment by all peoples of the highest possible level of health."

Index

Note: Page numbers followed by *f* and *t* indicate figures and tables, respectively.